Contaminated land

investigation, assessment and remediation

ICE design and practice guides

The First Edition of this guide was one of the most popular publications of the Institution of Civil engineers when it first came out in 1994. The purpose of the guide, as with the first edition, is to provide an introduction to the main principles and important aspects of the particular subject, and to offer guidance to the non specialist as to appropriate sources for more detailed information. The guide's aim is to supplement and enhance existing authoritative and competent guidance documents prepared by key organisations such as central government, the Environment Agency, the Construction Industry Research and Information Association (CIRIA) and the Building Research Establishment (BRE). It recognises the whole area of contaminated land investigation and remediation has undergone very significant changes since 1994 and will continue to do so for the foreseeable future.

The Institution continues to target as its principal audience practising engineers who are not expert in or familiar with the subject matter. This group includes recently graduated engineers who are undergoing their professional training and more experienced engineers whose work experience has not previously led them into the subject area in any detail. In an environment where new regulation and its interpretation is the norm, those professionals who are more familiar with the subject may also find the guide of value as a handy overview or summary of the principal issues.

As before, the guide features checklists to be used as an aide-memoire on major aspects of the subject and refers to references, bibliographies, guidance on authoritative, relevant and up-to-date published documents and recognises that web references provide the opportunity to keep the document as live and relevant as any printed material can be.

ICE design and practice guide

Contaminated land

Investigation, assessment and remediation

2nd edition

Jo Strange and Nick Langdon

Published by Thomas Telford Publishing, Thomas Telford Ltd, 1 Heron Quay, London E14 4JD. www.thomastelford.com

Distributors for Thomas Telford books are
USA: ASCE Press, 1801 Alexander Bell Drive, Reston, VA 20191-4400, USA
Japan: Maruzen Co. Ltd, Book Department, 3–10 Nihonbashi 2-chome, Chuo-ku, Tokyo 103
Australia: DA Books and Journals, 648 Whitehorse Road, Mitcham 3132, Victoria

First published 1994, reprinted 1995, this edition 2008

A catalogue record for this book is available from the British Library

ISBN: 978-0-7277-3482-2

Typeset by Academic + Technical, Bristol
Printed and bound in Great Britain by MPG Books, Bodmin, Cornwall

Foreword

As predicted in the first edition, contaminated land has become the subject of some importance to civil engineers today as the need to reclaim and recycle land strengthens. Society demands sustainable solutions of engineers where growing pressure on land and natural resources dictate greater environmental awareness.

Since the launch of the First Edition, in which Mary Harris and Sue Herbert set down many of the principles to which engineers work, there have been significant changes of detail. A regulator for England and Wales, the Environment Agency has been formed from the National Rivers Authority with far greater powers and terms of reference than envisaged by the first document. Likewise in Scotland and Northern Ireland,[1] central regulators have been formed with wider responsibilities and stronger remits. Legislation, both national and European, has been enacted over the last few years and most local authorities are substantially into their programmes for developing contaminated land departments. It is important that professionals working in this subject area appreciate at the outset the necessity of involving experts in several different disciplines, since contamination involves many technical and non-technical issues. In addition to civil and geotechnical engineering, expertise will be required in chemistry, environmental science, geology, hydrogeology, industrial archaeology and project management, and may be needed in other subjects such as chemical or process engineering, ecology, biology, valuation and other financial matters and legislation. No one civil or geotechnical engineer can have the breadth of expertise that will allow them to address all aspects of a contaminated land project.

Contamination can affect the ground, water or air. This guide concentrates on land contamination, which includes the water environment. Contamination can occur through the presence of chemical substances, either in solid or liquid form, or from noxious and hazardous gases.

Certain aspects of land contamination are particularly specialised, such as radioactivity or some forms of military contamination such as explosives or chemical/biological weapons. These subjects are not addressed in the guide and expert advice must always be sought in this respect.

[1] The Scottish Environmental Protection Agency (SEPA) was formed in 1996 taking over responsibilities from HM Industrial Pollution Inspectorate, the River Purification Authorities and the Hazardous Waste Inspectorate in Scotland, and in Northern Ireland, the Department of Environment has overall responsibility for environmental protection.

Acknowledgements

This guide has been prepared by Jo Strange and Nick Langdon of Card Geotechnics Limited and they are grateful for the strong framework and sensible approach of the First Edition set out by Mary Harris and Sue Herbert that has allowed a second edition to be prepared with a relatively light hand.

Preface

The Environment Agency in England and Wales and its equivalents in Scotland and Northern Ireland have become the leading custodians of the environment in the UK, both protecting and improving it. In the context of contaminated land, their prime role is one of protecting watercourses and aquifers. Recent years have seen an expanded role for local authority environmental services that has led to the appointment of many Contaminated Land Officers, charged principally with protecting human health where soil contamination and soil-borne gas concentrations become determining factors. To the non-specialist, this interplay between two regulatory bodies can be a source of confusion and, on occasions, frustration.

Recent years have seen the enactment of legislation on the environment and the adoption of many of the principles of the European Landfill Directive. This has meant specialists, regulators and the generalist engineer, are grappling with interpretation of guidance frameworks, new levels of testing accuracy and increased public perception of environmental damage.

The guide intentionally continues only to provide an overview of what is a very complex subject and draws heavily on authoritative and comprehensive technical guidance documents prepared by other bodies.

The knowledge and practical experience of contaminated land has matured and developed in the UK during the last 12 years. Greenfield sites are increasingly seen as the rarity. A risk management framework continues to be the principal approach to much that is being done today.

The two-part approach of the First Edition has been deliberately maintained, although the authors recognise that the techniques involved in remediation are subject to development, refinement and economic pressures that can quickly make authoritative discussion look dated.

Part I of the guide sets the use of the investigation methods within a risk management context and highlights those aspects where different techniques or a different emphasis is needed to ensure that contamination is adequately addressed. The guide describes risk assessment as a means of evaluating the significance of identified contamination and the development of the Conceptual Site Model based on plausible source–pathway–receptor scenarios as part of the assessment process.

Within a risk management framework, remediation methods aim to reduce the identified risks to acceptable levels.

Part II of the guide addresses setting targets for remediation, and the overall selection process to determine the most appropriate remediation strategy to achieve them.

Alongside the traditional civil engineering techniques used for remediation in the last 12 years, there are now more innovative and less engineering based solutions available. The principle of breaking the link between source and receptor is still the most favoured approach but source treatments such as bioremediation and techniques to 'fix' contamination are now commercially available when once they were little more than 'bench tests'.

More than ever, dealing with contaminated land requires a multi-disciplinary approach. The use of appropriately experienced specialists is vital to achieving cost effective and technically sufficient investigation, assessment and remediation. Use of technical guidance documents such as this are no substitute for specialist expertise.

Contents

Part I. Investigation and assessment

1. Introduction

1.1 What is contaminated land?

The Contaminated Land (England) Regulations 2000 became law on 1 April 2000, giving effect to Part 2A of the Environmental Protection Act 1990 and providing for the first time a legal definition of contaminated land and a new regulatory regime for its identification and remediation. Similar legislation came into force in Scotland and Wales on 14 July 2000 and 1 July 2001 respectively and has been established for Northern Ireland in 2006, but has not yet been brought into operation.

For all issues relating to contaminated land, the definition of contaminated land, as compared to public perception or those of professionals not involved on a day to day basis. The definition for regulation is given in Box 1.1.

Box 1.1 Contaminated land definition

'Land which appears to the local authority ... to be in such a condition that, by reason of substances in, on or under the land that; *significant* harm is being caused or there is the *significant* possibility of such harm being caused to specified receptors (targets), or pollution of controlled waters is being, or is likely to be caused'.

This definition reflects the intended role of the Part 2A regime, which is to enable the identification and remediation of land on which contamination is causing unacceptable risks to human health or to the wider environment.

Note: the definition does not include all land that has been impacted by chemical contamination, even though such contamination may be relevant in the context of other legislation or redevelopment works.

The 2002 report, The State of Contaminated Land,[1] by the Environment Agency quotes an earlier estimate that there may be as many as 100 000 sites affected by contamination to some degree in England and Wales. It advised that between 5 and 20% of these may require action to ensure that unacceptable risks are minimised. 'Environmental Fact and Figures' on contaminated land, published on the Environment Agency website in June 2007, suggest there may be as many as 335 000 affected sites, of which 10% have been identified as being contaminated and 21 000 have been treated by way of the planning system.

It is important to be aware that not all 'elevated' levels of potential contaminants are associated with man's activities. Likewise, it is important to recognise regional background levels of some 'common' contaminants can exceed normal guidance ranges. This is typically the case for arsenic in the south-west of England. The British Geological Survey[2,3] and other publications[4] provide appropriate guidance background levels on some of the more common heavy metals. In addition, an extension to Part 2A to cover radioactive contamination has been issued.[5] This could impact those areas affected by processed 'natural' radionuclides, but does not apply to radon.

The implementation of Part 2A of the Environment Act 1990 has meant that no longer is investigation of potentially contaminated land solely considered in the context of land redevelopment. Action may be taken by the local authority or Environment Agency to investigate, assess and remediate contaminated land where required.

However, the Office of the Deputy Prime Minister (ODPM) and its successor, 'Communities and Local Government', have set brownfield development targets which have been regularly revisited. (PPS 3, November 2006, refers to a requirement for a minimum of 60% of new housing on previously developed land). The result has been arguably the principal driver in the UK to tackle brownfield site redevelopment. Associated with these targets is the implementation of PPS23,[6] which has also placed a greater onus on the planning procedures to address issues relating to contamination.

Furthermore, under the PPC Regulations,[7] investigation of contamination may be necessary to provide baseline conditions for monitoring impacts of licensed activities. The basis for assessing the risks associated with land in use or occupied, and the practical and time constraints on investigation and remediation, mean that views on risk and potential for harm are different to those applying to a typical redevelopment scheme.

It is important to be aware that sites adjacent, or close to these industries or activities may have been impacted by contamination through such processes as atmospheric deposition, dust blow and the migration of contaminants in surface and groundwater. The potential for contamination is therefore not confined to the direct use of a site. It may reflect the impact of neighbouring uses or activities.

Identifying former site use and the potential contaminating process and individual contaminants is the essential first stage. Therefore, it is vital that the desk study is undertaken as an informative and interpreted activity rather than a data collection and box ticking process. Identified process and potential contaminants are identified in both CLR8[8] and BS 10175.[9] The Department of the Environment published a series of industry profiles,[10] which are available on the DEFRA website.[11] Typical examples are given below:

Industries and activities known to be associated with contaminated land:

Asbestos manufacture and use
Chemical industries
Dockyards
Explosive manufacture
Gas and electricity supply industries
Iron and steel works
Metal smelting and refining
Metal treatment and finishing
Mining and extraction
Oil refining, distribution and storage
Paints and graphics
Pharmaceutical
Scrap processing industries
Sewage works and farms
Tanning and associated trades
Transport industries
Use of radioactive substances
Waste disposal operations
Wood preserving

In addition to industrial uses of the type detailed above, contamination is regularly associated with tanks for storing heating oil. Indeed, a number of Part 2A site determinations are a result of this scenario. Even innocuous terrace houses can be the site of contaminative 'local cottage industries' such as dentistry, where discarded mercury can be found. Even allotments or waste ground/ bonfires/burnt-out cars can be the source of significant contamination.

1.2 Why is contaminated land a potential problem?

The UK has a long industrial history and many sites have been damaged as a result of their former use. Physical damage may be evident in the form of unstable ground, poor drainage, underground obstacles, voids and shafts, and topographical irregularities. Such features may pose problems for redevelopment or affect the aesthetic value of a site. Some may represent significant physical hazards, e.g. unstable ground or voids susceptible to collapse. However, where land is contaminated, the main concern is that substances are present which may pose significant risks to human health or the environment due to their toxicological or other hazardous properties. Humans potentially at risk from contaminative substances may include investigation personnel and construction workers, as well as site users and the general public. Health screening of personnel in regular contact with contaminated sites is desirable from both the employees and employers point of view. With current targets for development on brownfield land, most engineers will encounter the issues of developing potentially contaminated land in their careers.

Hazardous substances and materials that may be encountered on contaminated sites include those listed in Table 1.1. As examples, the following documents give guidance based on UK practice, but specific reference may be required to particular site activities to focus onto key contaminants on a site:

- CLR 8
- CA R&D report 66[12]
- CIRIA 132 a guide for safe working on contaminated sites[13]
- DoE industry profiles

The following definitions form the 'trinity' of risk assessment for contaminated land investigation.

A ***hazard*** is a property or situation that has the potential to cause harm. Hazards may be chemical (e.g. the presence of a potentially carcinogenic substance), biological (presence of a pathological bacterium) or physical (accumulation of an explosive or flammable gas).

A ***risk*** is the probability that harm will occur.

Harm may be damage to human health, other living organisms, environmental quality, e.g. of air, water, etc., or physical structures such as buildings or services. The current definition of contaminated land refers to *significant* risk of *significant* harm.

In a wider context, risks may be considered in terms of damage to financial interests and assets. Thus, site owners may take steps to limit potential environmental liabilities and increase the commercial value of their assets by investigating and remediating contaminated sites where appropriate. Purchasers will be concerned to avoid exposure to the same liabilities (and associated costs) through the acquisition of contaminated land and property.

Table 1.1 Hazardous substances and materials that may be encountered on a contaminated site

General category	Common examples	Typical site
Soil-borne gases and vapours	Methane, carbon dioxide, carbon monoxide, hydrogen sulphide, volatile hydrocarbons such as benzene	Landfill
Flammable and explosive gases	Methane, petroleum hydrocarbons	Tank farm, fuel storage sites
Biological agents or pathogens	Anthrax spores, polio, tetanus, leptospirosis (Weil's disease)	Abattoirs, landfills, canals
Flammable liquids and solids	Fuel oils and solvents feedstocks intermediates and products	Chemical works, fuel storage facilities, petrol stations
Combustible materials	Coal residues, ash, timber, variety of domestic, commercial and industrial wastes	Coal mines, coal yards, timber yard
Materials liable to self-ignition	Paper, grain, sawdust – if present in large volume and sufficiently damp to initiate microbial degradation	Warehousing, farms
Corrosive substances	Acids and alkalis, reactive feedstocks, intermediates and products	Chemical works, factories
Zootoxic metals (and their salts)	Cadmium, lead, mercury, arsenic, beryllium, copper	Wood treatment works, metal and electronics factories
Other zootoxic chemicals	Pesticides, herbicides	Agricultural premises
Carcinogenic substances	Asbestos, arsenic, benzene, benzo(a)pyrene	Old buildings, industrial sites, petrol stations, power stations, gasworks
Allergenic substances and sensitizers	Nickel, chromium	Metal works, timber treatment, batteries
Substances causing skin damage	Acids, alkalis, phenols, solvents	Chemical works, factories
Phyto toxic metals	Copper, zinc, nickel, boron	Metal works, electronics
Reactive inorganic salts	Sulphate, cyanide, ammonium, sulphide	Gasworks, farms
Radioactive substances	Some hospital laboratory wastes, radium-contaminated objects and wastes, some mine ore wastes, some non-ferrous slags or phosphorus slags	Hospitals, research facilities, power stations
Physically hazardous materials	Glass, hypodermic syringes	Hospitals, landfills
Vermin	Rats, mice, cockroaches	Anywhere, domestic waste

Source: CIRIA SP 101-112[15]

Therefore, land contamination becomes an issue of concern in relation to the risks it poses, or the perceived risk. The presence of measurable concentrations of chemicals in the ground does not automatically indicate that there is a contamination problem. Contamination may not pose a risk, but it may be difficult to dispel negative perceptions.

The risks associated with contaminated land should always be assessed in terms of:

Source – the contaminant(s) or hazard presenting a degree of risk to human health or the environment;
Pathways – the route by which a hazard comes into contact with a target or receptor and
Receptors – the entity that could be harmed through contact with a hazard.

It is important to remember that a risk *does not exist* unless there is a plausible pollution linkage, i.e. a complete source–pathway–receptor relationship. The risk does not exist

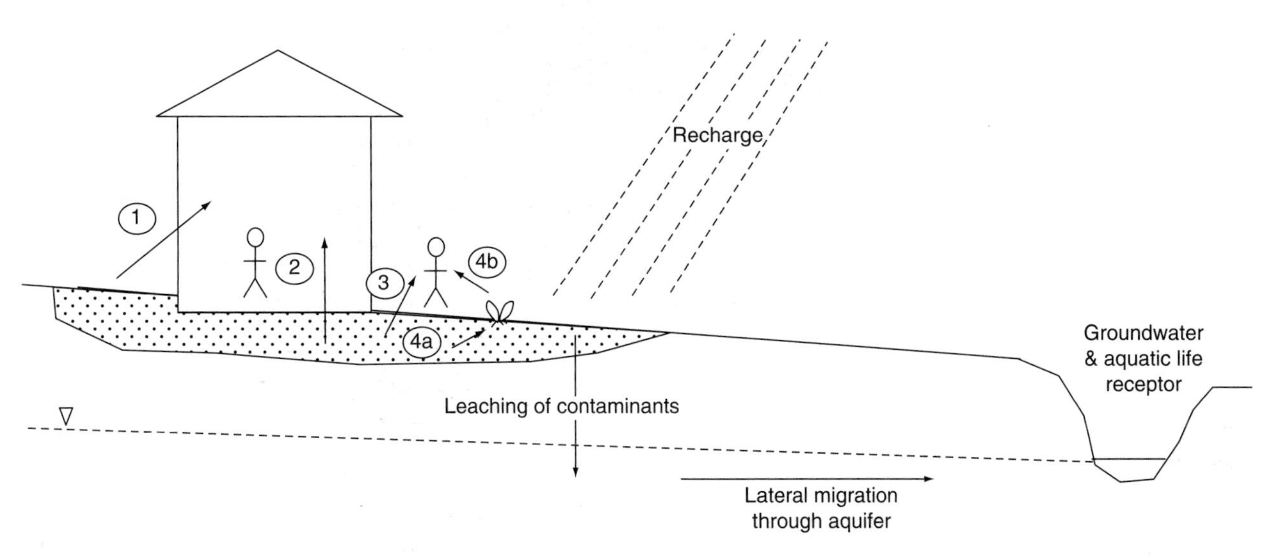

Figure 1.1 Example of conceptual site model

if any of the three links are missing or broken in some way. Examples of sources, pathways and receptors which may be relevant in contaminated land applications are shown in Figure 1.1. This presentation of the assessed potential risks associated with a site is referred to as a ***conceptual site model*** (CSM).

The degree of risk, and whether it is sufficiently serious to warrant action, depends primarily on the nature of the source–pathway–receptor relationship. Much of the practical effort of investigating, assessing and, where necessary, remediating contaminated land is therefore geared towards:

- identifying and characterising plausible source–pathway–receptor relationships
- establishing the nature and magnitude of the risks and associated effects
- deciding whether the risks are acceptable and, if they are not,
- deciding the best way of controlling or reducing the risks to an acceptable level, taking into account any practical, financial or other constraints
- planning, designing and implementing remedial action and demonstrating that it has been effective.

For contaminated land it is now accepted that this process is best handled within a formal risk management framework. Detailed guidance has been produced by the Environment Agency as CLR 11-Model Procedures for the Management of Contaminated Land, September 2004.[14]

1.3 Professional advice

Dealing with contaminated land requires the input of specialists with appropriate experience. These can be sourced from appropriate databases such as that of the Association of Geotechnical and Geo-environmental Specialists[16] (AGS) who require of their members a level of expertise, professionalism and commitment to training which, while present outside such a trade body, nevertheless are key requirements for those renewing membership each year.

Investigation, assessment and remediation of contamination involves a number of disciplines and will draw on professionals with backgrounds in applied geology, chemistry, ecology hydrogeology, toxicology, geotechnical engineering, and civil and environmental engineering who may well be described as geo-environmental specialists. Recently, professional bodies have seen the instituting of the Chartered Environmentalist, with designatory letters CEnv, and the SiLC (Specialist in Land Condition), administered by IEMA (Institute of Environmental Management and Assessment) and based on the use of Land Condition Records. Both professional qualifications are subject to submission and examination by peers in a manner not dissimilar to the Chartered Professional Review of the Institution of Civil Engineers. It is important that project managers for schemes which involve contaminated land appreciate and act on the need for specialist professionals to be part of the project team. The AGS is developing a guidance on the selection of suitable geo-specialists but, for general guidance, the International Federation of Consulting Engineers (FIDIC) has already published advice on selecting consultants.[17]

It is important to recognise the key role of the regulator in the whole process and the potential that the Contaminated Land Officer or Environmental Health Officer or the specialist within the Environment Agency may lack first hand experience of the construction industry or have specialised in construction orientated aspects of the broad spectrum that is contaminated land. Nevertheless, effective communication with the regulator to establish mutual understanding is a vital element of the process of investigating and remediating contaminated land.

Finally, and as a cautionary note, it is not the intention of this guide, nor indeed of other authoritative guidance documents, to provide an alternative to the use of appropriately experienced specialists.

2. The risk management framework

2.1 Scope

CLR11 defines 'risk management' as shown in Box 2.1.

Box 2.1 Definition of risk management

> 'Risk is a combination of the probability, or frequency, of occurrence of a defined hazard and the magnitude of the consequences of the occurrence.' The three essential elements to any risk are contaminant (i.e. source), pathway and receptor.

More recently the Institution of Civil Engineers has outlined the RAMPS[18] strategy which essentially looks to provide strategies in the event of a specific risk being realised. For contaminated land this might mean the identification of additional investigation techniques in the event of specific ground or groundwater contamination being encountered.

The main advantages of a risk management approach to contaminated land are that:

- it is systematic and objective
- it specifically provides for the assessment of uncertainty
- it provides a rational, consistent, transparent and defensible basis for discussion about a proposed course of action between the relevant parties (e.g. site owner, advisors, regulatory authorities, the local community).

There are four main elements:

- hazard identification and assessment
- risk estimation
- risk evaluation
- risk control/mitigation.

Hazard identification and assessment, risk estimation and risk evaluation together comprise risk assessment; risk evaluation and risk control/mitigation together comprise risk reduction.

The relationship between risk management and the main stages of a work programme of site investigation, assessment and remediation is shown in Figure 2.1. Although it is convenient to present the various components of risk management and associated

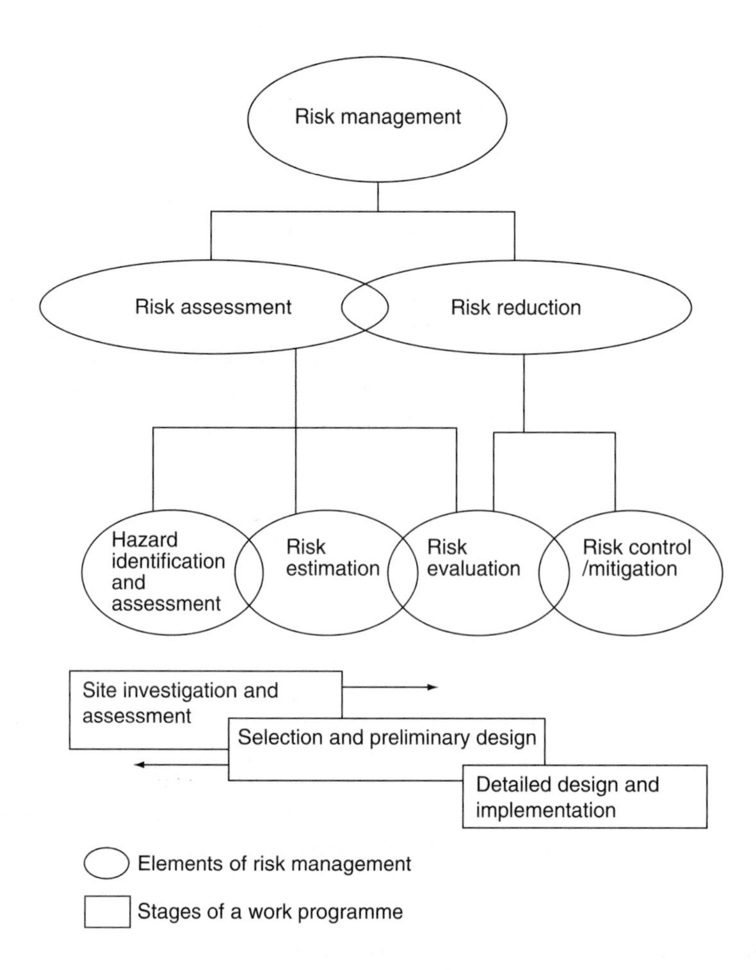

Figure 2.1 Relationship between risk management and the main stages of a work programme

work activities as discrete steps, there is considerable scope for iteration throughout the entire process. However, it is common for clients to look for single stages of investigation and cost prediction, even though this may result in greater overall expenditure. Iteration is a desirable feature of most contaminated land work because it allows:

– better targeting of effort
– more accurate definition of the problem and possible solutions
– better technical and financial control over all associated programmes of work.

2.2 Risk assessment

Risk assessment is that part of risk management which comprises appraisal of the significance of observed levels of contamination on a site. This requires the contaminative sources and risks to be identified and characterised, and their significance evaluated with respect to the relevant receptors. Historically there were two main approaches.

(*a*) The use of hazard identification and assessment, where judgements are based on the outcome of a comparison between observed levels of contamination and generic reference data indicative of specific types and level of risk. This provides a measure of the potential for harm – no attempt is made to estimate the probability that harm will occur. The most relevant example was the use of the ICRCL[19] trigger concentration values for soils (withdrawn in December 2002) and Waste Management Paper 27[20] values for soil gas. This generally still applies to the initial assessment of risk associated with phytotoxic metals using the limits published in the Sludge Regulations[21] and preliminary assessment of water quality against drinking water standards[22] and Environmental Quality Standard (EQS).[23]

(*b*) The use of site-specific risk estimation and evaluation, where some attempt is made to estimate the probability that harm will occur under the individual circumstances being considered: risk estimates may be qualitative (e.g. the risk is high, medium or low) or quantitative (e.g. the risk of a defined level of harm is less than 1 in 10).[6] A defined level of harm, as set by the Health Protection Agency, forms the basis of the current approach adopted in deriving the Soil Guideline Values (SGVs)[24] or other Generic Assessment Criteria (GACs) using the Contaminated Land Exposure Assessment (CLEA) framework documents. However, the CLEA framework allows for both generic quantitative assessments using published values or those calculated direct from the CLEA model as well as more complex detailed quantitative assessments for site specific conditions based on appropriate modelling.

CLR 11 provides current guidance with respect to a staged approach which refers to the preliminary risk assessment (qualitative) as an aid to develop the initial conceptual model of the site and establish whether there are potentially unacceptable risks. This leads to a generic quantitative risk assessment, assessing the applicability of generic assessment criteria and associated assumptions, and establishing actual or potential unacceptable risks. Typically, GACs are conservative and are based on largely generic assumptions regarding characteristics and behaviours of sources, pathways and receptors, within defined ranges of conditions. If necessary, this may be followed by detailed quantitative risk assessment. This uses site specific assessment criteria (SSAC) based on site specific information on contaminant behaviour and impacts, and hence determine whether the risk is unacceptable.

2.3 Risk reduction

Risk reduction or mitigation is that part of risk management in which it is decided that observed levels of contamination pose unacceptable risks to defined targets and measures are required to actively intervene at some point on the source–pathway–receptor linkage. This may include one or more of the following:

- decisions to be made on the type of response needed to control or reduce risks to a defined level, such as breaking the pathway or removing or reducing the hazard;
- remediation strategy based on these actions to be set out as a methodology, developed and put into effect;
- use of monitoring procedures to ensure that both the short- and long-term objectives of remedial action are achieved and the process is validated.

There will usually be several remediation options available; each will have different implications in terms of:

- technical effectiveness (including over the long term) in controlling/ reducing risks to the required level;
- feasibility (in terms of meeting other technical objectives, such as engineering properties, availability of necessary skills, plant, support services, time, minimal environmental impacts, etc.);
- acceptability to relevant parties, specifically the regulatory authorities;
- cost, which may well be assessed as part of an option matrix to establish cost benefit.

For the redevelopment of a contaminated site in the UK, the risk reduction objectives will take the form of agreed remedial measures and/or targets/acceptability criteria, which, for human health risks, take into account the proposed use of the site. These are normally expressed as residual concentrations of contaminants in soil related to

Soil Guideline Values (SGVs) from CLEA documentation, where available. Alternatively or in addition, site specific target levels (SSTLs) or agreed 'in house' GAC concentrations derived from appropriate toxicological information and UK relevant and EA endorsed models or quality standards considered appropriate for UK practice, may be adopted. For groundwater, remedial targets can be based on current drinking water standards or Environmental Quality Standards, with site specific values derived where applicable using appropriate contaminant fate transport models.

Soil remediation measures will be more stringent where the site is required for residential or agricultural use as opposed, for example, to hard cover, commercial or industrial use. Current land use definitions as adopted by the CLEA framework relate to:

- residential with plant uptake
- residential without plant uptake
- commercial/industrial.

(Note: there is no current equivalent to the former ICRCL land use of parks, playing fields or open space. Further, there are currently no immediately appropriate levels for highway, or railway landscaping for which there is a need to consider the appropriateness of the derived risk assessment values for the published land usage. Professional judgement is required in adopting an appropriate GAC *or* SGV, *or a* SSAC *should be derived from first principles.)*

The phrase 'clean-up standards' continues to have common use. It is also commonly misleading since there is generally no formal definition of 'clean' and the risk assessment will show different residual concentrations of contaminants to be acceptable depending on end use and remediation measures adopted. Further, principles of sustainability and waste minimisation which society is looking for in engineering projects make the 'cleaning' of a site by complete removal of contaminated soils to another 'landfill' increasingly unacceptable both socially and in terms of environmental and financial cost. The use of acceptability criteria is more appropriate in the current context.

Other engineering objectives may also allow remediation objectives to be achieved by reducing risks. For example, reduction in site levels may remove contaminated soil, construction of basement car parking with appropriate venting for vehicles may remove risks due to soil gas, and impermeable surfacing for drainage may prevent infiltration and migration of contamination to groundwater. Equally, siting of soakaways, poor detailing of cavity venting and service trench construction can increase the risk to site users, making it vital that those dealing with contamination are an integral part of the design and construction teams.

Some re-evaluation of the basis for action or adjustment to certain aspects of the scheme will become necessary and the process can become iterative. For example, it may be necessary to:

- adopt a different use of the site thereby changing the potential pollution linkages and hence appropriate SGV/GACs to be used;
- allow more time for completion of remediation, important where methods of bioremediation or active removal of products within the groundwater occur;
- 'engineer out' the problem, adopt hard landscaping and planters, raise site levels, phase developments to allow localised remediation or reshape the development (not easy with planning constraints).

2.4 Communication aspects

The communication of risk-based information to third parties is an important element of all stages of risk management. It requires particular care to ensure that information is accurately and appropriately presented and that it results in well-informed and constructive interaction.

Knowing the target audience allows communications to be better tailored. Typically, communication will be primarily with the client and their funders with documentation also being conveyed to regulators and other construction professionals. As a result, relatively technical information may require greater explanation or re-presenting when it is being communicated to local communities, politicians or special interest groups.

Complex investigation information, risk assessments findings and technical proposals for remediation have to be prepared in a way that allows the chosen method of delivery, such as a report, presentation or public meeting, to be effective. In some circumstances, engineers can forget the essential message in efforts to bolster technical credibility by using repetitive facts, graphs and tables when a few crisp bulleted points suffice.

As discussed earlier, use of the terms 'clean' in contaminated land applications is undesirable when dealing with third parties. Debate on what is clean has in recent years been included in legal actions[25] with very significant costs awarded to third parties whose view on what constituted 'clean' compared to the current practice interpreted by construction professionals was supported by the courts.

Within the current planning regime, PPS23 requires applicants to undertake a desk-based study and establish a Conceptual Site Model to establish risk and potential remediation as part of a planning submission. Without it, the regulators will place over-arching conditions on any planning permission, or refuse the application. Both the Local Authority, through Planning, Building Control and Environmental Health, and the Environment Agency, who are statutory consultees to the planning process, will set conditions together with the Environment Agency who themselves are statutory consultees to the process. Depending on the nature of contamination and the associated risk either the Local Authority by way of the Environmental Health Officer (EHO) or Contaminated Land Officer (CLO) or the Environment Agency will take the lead role.

Box 2.2 Lines of responsibility for dealing with contamination risk

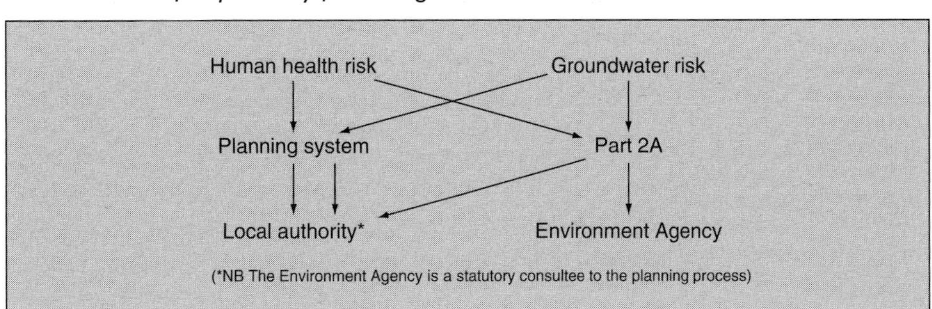

In simple terms, where risk to human health is a factor, such as heavy metal concentrations in soils, hydrocarbon vapours beneath completed residential developments or a pathogen, such as anthrax within soils placing construction workers at risk the EHO/CLO might be expected to take the lead. Where contamination in soils is likely to leach to controlled waters (both surface water bodies and groundwater), or contaminants are identified in site groundwater, the Environment Agency can be expected to take the lead. On some sites with complex contamination situations both bodies have roles taking leads in areas which to the non-specialist seem indistinct and confusing. On these occasions clients can believe approval from one body effectively removes the interest of the other. This is not so and the assumption causes confusion, frustration and delay.

Since 2004, increasing constraints on off-site disposal of waste, particularly hazardous waste, have resulted from the implementation of the European Landfill Directive.[26] Critical factors in the cost effectiveness of engineered solutions, such as off-site disposal of excess soils, have seen and will continue to see increases in cost and loss of appropriate disposal sites in the next few years. Detailed communication on waste classification between waste producers and the receiving facilities is now essential to ensure compliance with the relevant waste regulations.

Waste regulations also apply to the use of remediation processes and effective communication with the Waste Regulators within the EA is absolutely necessary if remediation schemes are to be approved and the necessary licenses issued for implementation.

3. Site investigation and risk management

3.1 Introduction

Site investigation and assessment are critical to the effective risk management of contaminated land. Their findings provide the basis for decisions on the need for, and type of, remediation. Ultimately, they are critical for the design and implementation of the necessary works and in their absence permission for houses to be occupied can be refused or works demolished. Deficiencies in site investigation and assessment lead inevitably to errors of judgement. These may affect the technical sufficiency, cost and duration of remediation for contaminated land projects as with all investigations for construction schemes.[27] Poorly informed and executed site investigation works may also expose investigation personnel and the general public to unacceptable health risks and could lead to more extensive or intractable contamination problems than those that previously existed on the site.

Figure 3.1 Night rotary drilling

Key to the success of any investigation is the desk study done *not as a document collecting exercise* or a *web down-load* but as a conscious interpretative process with the purpose of assessing risk and seeking methods to determine the reality of ground conditions. A multi-disciplinary approach is especially important for investigations into contamination, and project managers should ensure that the investigation team includes appropriately skilled professionals covering geology, chemistry, hydrogeology and geotechnics, together with other disciplines as needed for the particular scheme.

The commissioning of separate geotechnical and environmental investigations is neither cost effective nor desirable for most projects.[28] The purpose of any site investigation is to reduce the risk to client and third parties from ground conditions. This has been embodied in BRE guidance on brownfield sites.[29] Intrusive investigations will require the progression of boreholes, installation of monitoring equipment, recording of ground conditions and testing of soils. It is therefore illogical and extremely costly to a client to undertake, say, 15 m boreholes for environmental purposes, recorded to a lower and less recognised standard than BS 5930: 1999,[30] without in situ testing, only to 'later' require 20 m boreholes with SPTs and U100s for the engineering design of foundations. Likewise, hiring an excavator to sample to no more than 0.5 m across a site for contamination samples, when 3.5 m to 4.5 m deep trial pits would have benefited the civil engineering design and incidentally the contamination assessment is nonsensical but a frequent practice.

3.2 Site investigation

3.2.1 Purpose

The overall purpose of any site investigation is to ensure that an appropriate information base exists for assessing and managing the risks that may be associated with the ground and this includes contamination. It has become an industry axiom that 'clients always pay for a site investigation, even if they don't have one'.[31] Unfortunately, there are clients and their professional advisors who are still prepared to take the risk of ignoring ground conditions or indulging in 'wish fulfilment'. Since site investigation effectively underpins all the decisions and actions subsequently taken in connection with a contaminated site, it is essential that:

- it is properly designed and executed by appropriately skilled personnel and, to this end, bodies such as the AGS provide clear guidance;
- the intrusive investigation and monitoring is formulated on the basis of an appropriately comprehensive desk study;
- it provides information that supports the risk assessment and proves or otherwise the conceptual site model, i.e. it must address all relevant potential sources, pathways and receptors;
- it poses minimal risks to site personnel, the general public and the wider environment.

Investigation also requires access to the site and, in its later stages, typically involves physical disturbance of the ground. In some circumstances, site-based work may necessitate prior approval from the regulatory authorities. It is essential that this is in place before site work begins. This is particularly the case where investigation through contaminated soils and groundwater may see boreholes progressed into permeable, hitherto uncontaminated, soils of an aquifer. In these cases adoption of clean drilling techniques is likely to be the minimum constraint imposed by the Environment Agency. The consequences of unwittingly contaminating an aquifer

Figure 3.2 Cable percussion rig

are potentially extreme, and notification of the Environment Agency and submission of method statements on drilling techniques should be part of the investigation planning process.

Strictly, in England and Wales, notification is required to the Natural Environment Research Council (NERC)[32] under Section 198 and Section 30 of the Water Resources Act 1991, where boreholes deeper than 30 m and 15 m are proposed where water may be abstracted and under the Mining Industry Act for boreholes deeper than 30 m. It is also noted that any intrusive investigation adjacent (i.e. within 8 or 16 m depending on whether tidal or non-tidal conditions apply) to a water course will require a 'land drainage' permit from the EA.[33] Similar conditions are imposed by British Waterways for work adjacent to canals.[34]

Adequate investigation and assessment also require time, although client imperatives do not always recognise this. The lack of sufficient time during the planning stages of a project is one of the main factors limiting the effectiveness of the risk assessment process. Professional advisors should make every effort to explain to their clients any significance of time constraints with respect to the particular circumstances of an individual project.

3.2.2 Scope and objectives

Defining the scale and nature of any contamination present on the site is clearly a key objective of site investigation and will be based on the findings of a relevant desk study. However, the investigation is not sufficient in itself to permit an adequate assessment of risks. As indicated in Chapter 2, this requires that all potential pathways and receptors, including those associated with the off-site migration of contaminants, are properly identified and defined in the conceptual site model.

Table 3.1 Aspects to be considered during the investigation of contaminated sites

Aspect	Rationale
Contamination	Provides information on potential hazards; information requirements include nature (e.g. chemical/physical form, etc.), extent and distribution (on and off site) of the contaminants(s)
Geology	Provides data on the physical characteristics of the ground and of contaminated media: these may affect the location and behaviour of contaminants (e.g. adsorption onto clay/organic matter; migration by way of underground mineworkings) and the type of remedial action that can be taken (e.g. implications of underground obstacles, services, potential for subsidence during dewatering, etc.)
Hydrogeology	Provides information on sensitivity of potential water receptors (e.g. streams, rivers, lakes, groundwater) and likely transport and fate of contaminants in the water environment
Pathways and receptors	Fundamental in determining whether there is a risk of harm (i.e. whether plausible hazard–pathway–receptor scenarios exist), and the magnitude and severity of the risk(s) in relation to existing and likely future conditions
Geotechnical	Common requirement for intrusive investigation and or monitoring. Also increasingly required for quantitative risk assessment modelling

In addition to contamination aspects, the investigation must address the geological and hydrological properties of the site. These affect the behaviour of contaminants in the environment and may have a bearing on potential receptors (e.g. the water environment may provide a migratory pathway or itself represent a target at risk of harm; foreseeable events such as flooding could affect the distribution of contaminants and the targets at risk). By necessity, the investigation becomes more holistic in its approach identifying and determining factors that have relevance to the geotechnical and civil engineering solutions being sought by others in the design team. Geo-environmental and geotechnical investigations should not be seen as mutually exclusive but two sides of the same risk reducing operation undertaken for a client.

On occasion, off-site investigation work may appear to be appropriate, but this is typically unlikely to be possible, without regulatory intervention, as third parties have no incentive to allow access. The involvement of parties outside the development or direct site ownership mean that inevitably such elements of an investigation are costly, lead to delay and subject to the vagaries of partial advice and legal action.

The aim of any investigation where contamination is suspected should therefore be to address five main aspects (see Table 3.1), although others will exist associated with engineering solutions:

- identifying the contaminating substances
- establishing the local and regional geology (including geotechnical aspects)
- hydrology (including both surface and groundwater)
- identifying the potential natural pathways and receptors
- providing geotechnical design parameters.

An investigation strategy which addresses all five aspects on an integrated basis is a necessity and will lean heavily on the methodologies and procedures advocated in BS 10175: 2001 and CLR11.

Site investigation is also required to meet a wide range of other practical information needs. Typically these may include establishing the health and safety requirements of investigation and remediation personnel or identifying any factors that may affect the feasibility of undertaking particular types of remediation. Specific objectives such as those listed below should be developed to assist in the design and implementation of site investigation work.

- To determine the nature and extent of any contamination of soils and groundwater on the site and soil-borne gasses.
- To determine the nature and extent of any contamination migrating off the site into neighbouring soils and groundwater.
- To determine the nature and engineering implications of other hazards and features on the site (e.g. expansive slags, combustibility, deep foundations, storage tanks, sulphates).
- To identify, characterise and assess potential receptors and likely pathways.
- To provide sufficient information (including a reference level to judge effectiveness) to identify and evaluate alternative remedial strategies.
- To determine the need for, and scope of, both short- and long-term monitoring and maintenance.
- To formulate safe site-working practices and ensure effective protection of the environment during remediation.
- To identify and plan for immediate human health and environmental protection and contingencies for any emergency action.

3.2.3 Main phases of investigation

At the outset, it is clearly not economic or feasible to examine in detail all areas of a site or to test for all possible contaminants. However, it is noted that client pressure may force a single intrusive investigation process on the design team. Nevertheless, at the beginning of an investigation it is rarely apparent what all the main priorities should be. Specific health and safety, and environmental protection requirements, are also unlikely to be known with certainty. Phasing is encouraged and BS 10175 offers a valuable means of identifying and refining site investigation priorities, data gathering and evaluation.

The investigation of contaminated sites should involve at least *three* phases (preliminary or desk-based investigation coupled with site walk over, a detailed intrusive investigation and validation or compliance/performance investigations). Some investigation may see up to five phases:

- Phase I Investigation – preliminary investigation (comprising desk study and site reconnaissance).
- Phase IIa Investigation – exploratory investigation (e.g. preliminary sampling, monitoring).
- Phase IIb Investigation – main investigation (involving detailed on-site exploratory work).
- Phase IIc Investigation – supplementary investigations (the collection of additional site investigation data for specified purposes).
- Phase III Validation and testing for compliance and performance (comprising on-going monitoring and validation of remedial action, and post-treatment management).

There is scope for overlap between these phases. For example, some initial exploratory work (e.g. sampling and analysis of surface deposits) may be combined with site reconnaissance; supplementary investigation may form the final stage of an

Table 3.2 Examples of objectives and activities associated with site investigation

Phases of investigation	Typical objectives	Typical activities
Preliminary investigation	To provide background information on past and current uses, hazards, geology and hydrology, possible scale of contamination, etc. To inform design of on-site work (including sampling and analysis, health and safety, environmental protection) Can be used to rank a number of sites based on hazard potential May provide initial indication of remediation requirements	Literature review Consultation (e.g. site owners, neighbours, regulatory authorities) Site visits
Exploratory investigation – only required in relevant circumstances – not in all cases	To confirm initial hypotheses about contamination and site characteristics To refine design of detailed investigation	Preliminary sampling and monitoring, e.g. surface soils, water and/or vegetation
Main investigation	To fully characterise contaminants, geology, hydrology of site and associated pathways and receptors To inform risk assessment and selection of remedial methods	Comprehensive investigation of ground (e.g. using trial pits, trenches, boreholes) Monitoring (e.g. gas composition and water quality, flora and fauna)
Supplementary investigations	To obtain additional information in support of risk assessment and/or selection of remedial strategies	Further ground investigation and monitoring Treatability testing
Validation testing for compliance and performance	To confirm effectiveness of remedial action	Post-treatment validation and monitoring as appropriate

extended main site investigation. However, site investigation personnel should not enter a site, and no initial exploratory work should be carried out, *unless and until a desk study has indicated that it is safe to do so.*

The desk study will also provide the information necessary to determine appropriate health and safety precautions for the investigation personnel undertaking subsequent phases. For example, the British Drilling Association/Institution of Civil Engineers Site Investigation Steering Group classification system, using green, yellow and red for potentially contaminated sites for drilling purposes,[35] should be referred to.

In addition, certain circumstances will require specific licenses to dig or investigate, such as in close proximity of canals or river banks, where anchored river walls exist or strategic services occur and their procurement forms a critical part of the desk-based studies.

Phasing can allow shallow depth contamination to be evaluated before deeper, perhaps uncontaminated, horizons or aquifers are penetrated or an overview gained, sufficient for, say, site purchase.

Each phase in the investigation of a contaminated site probably has a different set of objectives, and involves different types of activity (see Table 3.2). It is important to note that the findings of each phase are used as the basis for designing the next. Thus, a preliminary site investigation report should set out the conclusions drawn from the work, and a series of objectives and plans for the next phase, if required. Such recommendations must be realistic and not an excuse for 'job creation'.

For economic and data interpretation reasons, the aim should be to conduct an integrated investigation so that all four main aspects of contamination, geology, hydrology and pathways/receptors can be addressed at the same time. However, the extent to which this can be achieved in practice varies depending on the phase of the investigation.

Integration of the preliminary investigation should pose the least difficulty, although different reference data should be consulted during the desk study (see Appendix A for typical information sources on contamination and hydrological aspects). However, during the main and supplementary phases of site investigation, specific provision (e.g. siting of boreholes, wells, trial pits, numbers and types of samples collected, etc.) may have to be made to gather information on different aspects, for example for geotechnical purposes as well as contamination objectives. It is always important to remember that these objectives are not mutually exclusive. Conflicts can occur, but compromises are usually possible with communication between elements of the design team. For this reason, it is important for those undertaking the environmental investigation to have a clear understanding of construction processes, if not the detail, and civil and structural engineers have understanding of the timescales for testing and monitoring and need to install and maintain monitoring positions through the construction process.

A wide range of techniques is available for characterising the contamination, geological, and hydrological profile of a contaminated site. Each has advantages and limitations that must be addressed on a site-specific basis at an early stage in the design of an investigation. (See Appendix B for further information on available

Figure 3.3 Window/sampler rig

Figure 3.4 Trial pitting

methods and applications.) For all activities, the 'designer' of the investigation has responsibilities under the Construction (Design and Management) Regulations 2007[36] for safe design and operational practice.

Different techniques have different implications for:

- the ability to install monitoring equipment
- the ability to physically access the site
- the quality and interpretation of site investigation data
- the health and safety of site personnel and the general public
- environmental quality.

Some examples are:

- Access may restrict equipment to lightweight window sampling, limiting investigation to depths no greater than about 10 m.
- Rotary drilling, particularly using air flushing, in contaminated ground may affect the distribution of volatile contaminants in the formation being sampled and may pose potential hazards to the workforce and general public.
- Trial pitting may release large areas of soils that are effectively capping contamination, hence releasing volatile gases into the local environment or providing a new pathway for surface water to leach contaminants.
- The failure to take appropriate precautions when sinking a borehole to significant depth at a site overlying a sensitive aquifer, where potentially mobile contaminants are present, is likely to make an already difficult problem considerably worse.

Appropriate techniques must be selected on the basis of the objectives of the particular phase of the investigation, the geological and hydrogeological conditions and the known or suspected contaminants present. For example, very different techniques could be used for the investigation of potentially combustible material compared with a site thought to contain significant concentrations of heavy metals.

3.2.4 Sampling and analysis

Sampling, analysis and on-site testing strategies must be developed for:

- initial exploratory investigations
- main site investigations
- supplementary investigations
- validation and monitoring programmes.

Box 3.1 Typical issues to be addressed when developing sampling and analysis strategies

Sampling

- What types of samples should be collected (e.g. soils, waters, wastes, gas/air, vegetation)?
- How many samples should be collected (across the site, what depth and where)?
- How much sample should be collected?
- How often should samples be collected (over time)?
- How should samples be collected?
- How should samples be stored?
- How should samples be transported?

Analysis and on-site testing

- What types of analysis/testing should be conducted (e.g. chemical, biological, physical)?
- How many datasets are required to be statistically viable?
- How should samples be prepared for analysis (e.g. no preparation, drying, grinding, sieving)?
- What level of detection is required?
- What level of precision is necessary?
- What analytical testing techniques should be used?
- How should the data be reported?
- How quickly are the data needed?

General

- What quality assurance/quality control procedures should be applied to ensure the validity of the results (e.g. use of MCERTS/UKAS accredited laboratories)?[37]
- Material certification.

It is clearly not possible to collect an infinite number of samples or to test for unlimited numbers of contaminants. Since assessment of risk is the strategy framework, exhaustive testing is not expected. In keeping with the overall approach to site investigation, sampling, testing and analysis strategies must be developed according to the specific objectives of each phase, taking into account what is already known about the site. Care must be taken to avoid a false sense of certainty imbued by reliance on theoretical sampling patterns. These assume homogeneity of soil properties such as permeability and soil density, lack of local man-made irregularities, the uniformity of contaminant as a neatly described plume, and a large client purse. All of which are usually very far from reality. Sensible interpretation of desk-based material combined with site observation is likely to provide a greater degree of confidence in the testing whose purpose is to allow risk reduction. Typical issues to be addressed when developing sampling and analysis strategies are listed in Box 3.1. More detailed information and guidance is available in BS 10175, BS 5930, CLR 4[38] and CIRIA SP103.[39]

A phased approach to sampling and analysis offers similar benefits to a phased approach to site investigation as a whole. However, it has the downside that delays in testing certain contaminants, particularly volatile hydrocarbons, will potentially result in deterioration of samples and misleading results, particularly if samples are not stored properly at all times. However, a phased approach allows for the gradual accumulation of information on the types and location of contaminants present on the site, and the progressive refinement of sampling and analysis objectives and procedures. There are several ways in which phasing can be achieved in practice (see Box 3.2). Further refinements, focusing, for example, on particular areas or materials of concern, or the use of more sensitive, substance-specific analyses, can then be made on the basis of initial findings.

Box 3.2 Examples of a phased approach to sampling and analysis

- Initial requirements can be established by desk study of the past and current uses of the site; the types of materials (e.g. as raw materials, products, by-products and wastes) stored, used or otherwise handled on the site; and details on the location of process plant, waste disposal and storage areas etc. Reference can be made to 'typical' contaminant lists such as those that exist in CLR8 and BS 10175. It is important to recognise that these are extremely comprehensive listings and as such some selection of testing class types of contaminants or discussion with the analytical laboratory is advisable.
- Background information on the geological/hydrological characteristics of the site, combined with data on the likely behaviour of contaminants in the environment, can be used as a guide to sampling locations/frequencies.
- Field observation (e.g. visual and olfactory) and on-site testing (e.g. by photo-ionization detector (PID) or test kit) can be used to select appropriate samples for analysis.
- Samples collected in the field can be retained for analysis at a later date (provided this is compatible with storage and sample preparation constraints).
- Analytical screening techniques, such as inductively coupled plasma (ICP) and gas chromatography/mass spectroscopy (GCMS) can also be used to provide a broad indication of the types and approximate quantities of different substances present on the site.

The development of sampling and analysis strategies is highly site-specific and is still to a large extent guided by the experience and specialist expertise of site investigators and laboratory personnel. While 'judgemental' sampling and analysis is extremely valuable, in terms of increasing the efficiency and minimising the cost of site investigation, it should not be regarded as a substitute for a thorough and sound approach that ensures statistically justifiable amounts of data.

Guidance on the statistical basis for sampling contaminated land was published by CIRIA[39] in 1995. This suggested the importance of some form of systematic grid for locating sampling positions across a site. It also recommended that a statistically based regular grid sampling was useful to define the lateral boundaries of contamination (such as a 'hot spot') with a specified degree of confidence. While academically and statistically justifiable, for the vast majority of sites, the cost of the implied testing is prohibitive and the need to define boundaries of hot spots in this manner unjustifiable when excavation and in situ testing are in any case required for the validation process. It is vital to remember that contaminant distributions usually vary significantly horizontally and with depth. Sampling positions and frequencies must be sufficient to define the vertical extent of contamination as well as the lateral extent (e.g. by collecting samples below the level at which visual or other information on the anticipated behaviour of the contaminant suggests contamination is unlikely to be present). The selection of grid size must be a matter for professional judgement based on the site characteristics and the investigation objectives. CLR 7 allows for 'averaging areas' based on proposed land use and statistical assessment of data from such areas.

In general, it is easier to show that a site is contaminated than to prove that it is uncontaminated. Whatever sampling strategy is adopted, therefore, it must be sufficient to demonstrate unequivocally which parts of the site are unaffected (and can therefore be safely excluded from any remedial programme) even if this means extending the sampling and analysis programme beyond that required, simply to show the presence of contamination. The current risk-based approach of CLEA documentation allows this through the consideration of the 95th percentile criteria. Unfortunately, the absence of toxicology data relevant to the United Kingdom to underpin the SGVs recommended for some contaminants has hindered a comprehensive roll out of advice from the regulators leaving commercial organisations to derive in-house Generic Assessment Criteria (GACs) for some contaminants. However, DEFRA has planned a way forward to address many of the issues relating to lack of and inappropriateness of some SGVs, which is detailed in a 2006 advice note subtitled 'Soil Guidance Values: the Way Forward'.[40]

It is also important to understand the construction process itself. For example, while theoretically the contamination can be assessed by sampling in each domestic garden of a residential development; on a contaminated site it is more likely that a relatively few samples are sufficient to establish the need for remedial capping. The critical sampling and testing then becomes that of the imported topsoil and subsoils to show them to be acceptable. Since these soils are often being imported by contractors from other development sites it is not unknown for the imported soils to contain potential contaminants and on site testing of topsoil batches/deliveries becomes the critical validation process.

3.2.5 Legal aspects

Depending on the exact circumstances, site investigation works may be subject to prior authorisation by the regulatory authorities under Part 2A or PPS 23. Other requirements may have to be satisfied by virtue of the health and safety implications of the work.

Table 3.3 gives brief guidance on the legal provisions that may apply on a site-specific basis. More detailed information on the legislative framework applying to contaminated land, present and past, can be found in the *Pollution Handbook*[41] and CIRIA SP112.[42] Note that the guidance in Table 3.3 applies to the situation in England and Wales (different provisions apply in Scotland and Northern Ireland in some cases, as detailed in reference 2) and that the legislative framework may change from time to time. Appropriate legal advice should always be sought in relation to individual sites.

In addition, where relatively deep boreholes (in excess of 40 to 50 m) are contemplated or work close to the top of an aquifer, it is advisable to notify the Environment Agency of these intrusive investigations and be prepared to submit method statements for this element of the work.

3.2.6 Health and safety

On-site investigation work (including site reconnaissance) may expose personnel to health and safety risks. CIRIA Report 132 should be consulted. Hazards may relate

Table 3.3 Legal provisions that may apply to site investigation in England and Wales

Area of law	Legal provision	Requirement
Land-use planning and development control	Town and Country Planning Act 1990 Planning and Policy Statement 23 – Planning and Pollution Control, November 2004	Permission for development (which may include site investigation/monitoring operations in some circumstances)
Public health	Control of Pollution Act, 1974 Environmental Protection Act 1990 (EPA) Occupiers' Liability Act 1957	Obligation to prevent the creation of a statutory nuisance (e.g. generation of toxic vapours, noise, dusts, etc.). Obligation to ensure the safety of visitors (which may include trespassers) to premises
Determination of contaminated land	Part 2A EPA	Obligation to prove whether significant linkage exists
Health and safety	Health and Safety at Work, etc. Act 1974 and associated regulations	Safety of employees and the general public from hazards arising at a place of work
	Construction (Design and Management) CDM Regulations 1994 CDM Regulations 2006 (to be implemented Spring 2007)	Obligation on designers, client and contractors to ensure design and site operations can be undertaken safely. Appointment of Planning Supervisor (or Coordinator) when appropriate
Environmental protection:		
Air	Control of Pollution Act 1974	Powers to local authorities to make enquiries about air pollution from any premises, except private dwellings
Water	Water Resources Act 1991	Prior authorisation required from the EA to make a discharge of polluting substances to a controlled water
	Pollution Prevention and Control Act, 1999	Powers to EA to protect the aqueous environment and to remedy or forestall pollution of controlled waters
		Investigation to determine baseline conditions or conditions at permit closure
	Water Industry Act 1991	Prior authorisation required from the sewerage undertaker to make a discharge of polluting material to a sewer
Waste	Environmental Protection Act 1990	Duty of Care on all those involved in the production, handling and disposal of contaminated waste (i.e. spoil arising *from* trial pit/borehole) to ensure that they follow safe, authorised and properly documented procedures and practices
Protected areas, species and artifacts	Town and Country Planning Act, 1990 Wildlife and Countryside Act, 1981 Ancient Monuments and Archaeological Areas Act, 1979	Protection of designated areas (e.g. sites of special scientific interest), species (e.g. plants and animals) and artefacts (e.g. ancient monuments)

Figure 3.5 Floating rig

to substances (solids, liquids, gases) present on the site (above and below ground) or to its physical condition (e.g. unsound buildings or other structures, voids, unstable ground, etc.). Other risks such as Leptospirosis and other biological hazards may also have to be considered. Physical injuries (e.g. cuts, grazes) may enhance the risks associated with exposure to hazardous substances by creating a ready means of access into the body. The possible significance of health and safety issues is much greater for an investigation into potentially contaminated land than for a conventional geotechnical investigation. However, the HSE in recent times has taken a keen interest in the safety of operation of drilling equipment, regardless of purpose.

A site-specific risk assessment should always be done prior to any member of the team going to site, to identify hazards that can be as diverse as concentrated solvents, or syringes to lone working and unstable ground/building. The development of appropriate health and safety provision is therefore a vital aspect of the design of any investigation of a contaminated site, and site reconnaissance or inspection visits should not be made without proper consideration of the risks to the personnel involved or to the general public.

In addition to the Health and Safety at Work Act, 1974, site investigation work is subject to other health and safety legislation including the Control of Substances Hazardous to Health (COSHH) Regulations, 1988. These require an assessment to be made of all relevant health risks before a work activity commences, and the use of appropriate control measures where necessary. The Management of Health and Safety at Work, etc. Regulations 1992, extend the assessment to cover all types of hazards and hazardous activities, and include the welfare of the general public. The Construction (Design and Management) Regulations 2007 require consultants and specifiers to ensure that their designs, which includes the investigatory process as well as remediation works, can be undertaken safely and that statements of working methods are adhered to. Processes must also be in place to deal with accidents and emergencies. This is of particular importance where decontamination units are to be present on red or yellow sites or where major services exist. It is not unusual for so-called decommissioned services to prove to be anything but or for 'unmapped' ducts or pipes to be potential containers of contamination.

Specific requirements will vary depending on the nature of the site and the phase of investigation. Typical issues to be addressed when developing health and safety plans are listed in Box 3.3.

Box 3.3 Health and safety issues

Health and safety procedures	Health and safety equipment
• Controlled entry (permit to work) procedures where applicable) • Site zoning (i.e. dirty and clean areas) • Good hygiene (e.g. no smoking, eating except in designated areas) • Monitoring (e.g. for on-, off-site toxic/ hazardous gases) • Appropriate disposal of wastes • Safe handling, storage and transport of hazardous samples • Control of nuisance (e.g. noise vibration, dust and odour) • Emergency procedures • Provision of appropriate training (e.g. to recognise hazards, use equipment) • Need for routine health surveillance	• Washing and eating facilities • Protective clothing (e.g. for eyes, head, hands and feet) • Monitoring equipment (e.g. personal exposure, ambient concentrations) • Respiratory equipment • First aid box • Mobile telephone • Decontamination facilities (e.g. for boots, clothing, machinery)

More detailed information and guidance on health and safety provision during the investigation of contaminated sites is available in CIRIA 132.

3.2.7 Quality assurance and control

Quality assurance/quality control (QA/QC) in site investigation and assessment is an important means of confirming the validity of the procedures and data used for risk assessment purposes. All aspects of site investigation and assessment should be subject to QA/QC procedures. Frameworks which allow ISO 9001[43] and ISO 14001[44] accreditation for companies should allow the basic procedures to be present to achieve this. Critical to the whole process is not a box-ticking approach but a habit of mind within individuals and organisations to ensure experienced staff review, verify and appraise information and advice before issue to third parties.

Some typical activities where review and approval by senior staff can be critical are listed below. The aim should be to avoid the 'switch on report template, disengage brain' attitude that can become the customary practice for less experienced, and indeed experienced, staff.

- Identification of appropriately qualified staff for investigation
- Critical assessment of desk study documentary evidence
- Implementation of health and safety procedures – safe working and method statements
- Scope of site reconnaissance
- Critically appraise site conceptual model – potential source–pathway–receptor relationships
- Exploratory hole locations, depths sampling testing and monitoring arrangements
- Data recording, sampling and monitoring procedures

– Duty of care procedures for collection, handling storage and preparation of samples – laboratory suitability – MCERTS, United Kingdom Accreditation Service (UKAS) etc.
– Reporting of data in factual and interpretative reports
– Input to, and use of any models to aid interpretation of the data
– Compliance with all relevant legal requirements
– Establishment and performance of environmental protection measures
– Waste disposal arrangements (duty of care, etc.).

More detailed information on the use of QA/QC procedures in contaminated land applications, and in sampling and analysis in particular, can be found in EA Technical Report P5-066[45] and BS ISO 10381.[46]

3.3 Risk assessment

3.3.1 Objectives

In simple terms, the purpose of the risk assessment approach by the CLR 11 framework documentation is to determine:

– whether observed levels of contamination on a site are likely to pose unacceptable risks to defined targets now or in the future
– whether measures should be taken to reduce/control risks to an acceptable level.

Additional elements may become significant with individual sites, such as risk of inundation of a contaminated site in a 1 in 100 year flood and so on.

3.3.2 Hazard identification and assessment

Source or hazard identification and assessment involves collecting sufficient information about the contaminants, the site (including its geotechnical and hydrological characteristics), and the wider environment to identify, characterise and assess the importance of sources, pathways and receptors. Hazard identification and assessment is informed by site investigation, initially through desk studies and site reconnaissance, and subsequently through the main ground investigation or intrusive work.

Deciding whether observed levels of soil contamination are significant in terms of anticipated pathways and receptors typically involves the activities listed in Box 3.4. Assessment is required of each separate contaminant observed at the site.

Box 3.4 Hazard identification and assessment

Assessment of observed concentrations in accordance with the statistical approach required by CLR 7 to determine the Us_{95} (the mean at the 95th percentile) for comparison against published SGVs for appropriate site use. (The scope of this approach is still limited by the relatively few number of published SGVs. This is gradually increasing over time, as more work on this is carried out by the EA/DEFRA.) Comparison can also be carried out against generic assessment criteria (GAC), which are derived using the CLEA model.

Where SGVs are unavailable and GACs are not considered appropriate, derive appropriate SSAC (site specific assessment criteria) based on best available UK relevant toxicological data and using the most appropriate human health risk model. Current UK models include CLEA and SNIFFER.[47] Other model options may be used with care in appropriate circumstances and if configured to UK defaults. In this case, the modeller has to fully justify use of the model and demonstrate its applicability.

In the absence of appropriate data or a suitable model, reference is occasionally made to published values such as Dutch intervention values[48] and Canadian (CCME)[49] values. As with the UK models, these are also derived from toxicological human health risk modelling.

(NB: Care is required in application of such non-UK standards and their applicability and relevance to the UK geology and human health risk model should be carefully considered before use. For sensitive sites, use of non-UK assessment criteria and their acceptability should be established with the local authority acting as the regulator for specific applications.)

Where appropriate, refer to published UK guidance limits such as Sludge (Use in Agriculture) Regulations 1989 to assess phytotoxic risks.[21]

Review criteria against site conceptual model.

Check potential for naturally occurring background levels for some contaminants.

ICRCL 59/83, published by the UK Interdepartmental Committee on the Redevelopment of Contaminated Land, was used for limit concentrations for assessment of soil contamination. However, these values were only strictly relevant to coal carbonisation sites despite their widespread use. They were withdrawn in December 2002 shortly after publication of the CLEA framework. (Other ICRCL documents, provide useful guidance on contamination assessment, are still valid.)

At the time of writing, the adoption and suitability of SGVs for the assessment of historic contamination where sites are being determined under Part 2A of the Environmental Protection Act is the subject of considerable debate. However, it is apparent that the use of SGVs within this determination process ignores the differing levels of perceived risk raised by developing a site for residential use and assessing a residential development of some historic standing on a former industrial site as part of a decision process to determine the site.

For assessing ground and surface water contamination on brownfield sites, there are no specific guidelines. Assessment is usually carried out against published water quality standards such as the drinking water standards but also, where appropriate, depending on the sensitivity of the affected water body, adopting values for bathing water and Environmental Quality Standards for rivers. In the absence of relevant UK data, other national brownfield groundwater assessment criteria, such as Dutch, are sometimes adopted, but these should be used with care. Where critical, assessment values should be agreed with the Environment Agency and, if appropriate, groundwater contamination fate transfer modelling carried out to assess the significance of the measured contamination.

UK models such as R&D P20[50] and ConSim,[51] developed by the Environment Agency, the latter in conjunction with Golder Associates, are commonly used for this purpose. Although other more complex models and software do also exist, it should be recognised that the quality of the modelling will only be as good as the quality of the input data and the skill of the modeller.

In terms of assessing soil gases, WMP 27 has been superseded by TGN03[52] and CIRIA 149[53] along with the subsequently published work by Wilson and Card[54], uses combined effects of gas concentrations and flows to assess risks due to soil gases. CIRIA 152[55] describes a development-specific approach to quantitative risk assessment for gases. The risk-based approach to assessing soil gases is also embedded in the 2004 revisions to Part C of the Building Regulations.[56] CIRIA report C659,[57] issued in late 2006, summarises the latest guidance on gas assessment incorporating specific National House Builders' Confederation (NHBC) requirements.

Depending on the outcome, the hazard assessment may indicate that:

- observed levels of contamination are unlikely to pose a risk to specified receptors and no further action is required
- further investigation and/or assessment (perhaps involving site-specific risk estimation) is needed before the significance of observed levels of contamination can be properly judged
- levels of contamination are such that there is no doubt as to the need for remediation.

It should be borne in mind that the acceptability of the contamination is judged solely against the assumptions built into the generic guideline or standard being used. It is fundamentally important that the applicability of these assumptions is assessed for each specific circumstance in order to ensure that the outcome of the hazard assessment is defensible.

3.3.3 Risk estimation

Risk estimation involves detailed evaluation of sources, pathways and receptors to determine:

- the nature of the exposure of the receptor to the source
- the nature of the effects produced under defined levels of exposure
- the probability (expressed in either qualitative or quantitative terms) that adverse effects will occur under defined conditions of exposure.

Depending on the reference data used, the output of risk estimation may be expressed in qualitative terms, i.e. a narrative statement that the risk of a defined level of harm is high, medium or low, or numeric terms, e.g. the risk of excess cancer over the lifetime of the individual is less than 1 in 10^6.

Whether a qualitative or quantitative risk estimation is undertaken, it will usually involve the manipulation of quantified data describing the movement of the contaminant from the source to the target.

Numerical evaluation of risk is only possible where quantitative risk assessment has been carried out using some sort of modelling. The validity of the modelling relates directly to the accuracy and quality of the data used to produce the site conceptual model. The model defines the conditions which influence the behaviour of the contaminants through the pollutant linkages being assessed. Typically the model should include:

- chemical form and physical properties of the contaminant
- characteristics of the host medium (soils, rock, groundwater, etc.) and effect on contaminant concentrations along travel pathways

- concentration of contaminants at the source, at points along the travel pathway and at the point of exposure (e.g. ingested by the target)
- rate of movement along the pathway
- amounts, frequency and duration of exposure
- characteristics of exposure route (e.g. ingestion, inhalation, direct contact) that determine how much of the contaminant is taken in by the target
- data limitations.

Modelling of human health risk in soils requires toxicological assessment. The purpose of toxicological assessment is to determine the effect (e.g. toxicological, carcinogenic, mutagenic, corrosive, etc.) of the source on the receptor under the conditions of exposure defined in the site conceptual model. Effects assessments involve a consideration of:

- dose-response relationships, and in particular the nature of the response at no observable effect level (NOEL)
- biological mechanisms regulating responses to different types of substances
- factors affecting response of targets (e.g. gender, age, general health status, species composition, physical properties of the building fabric, etc.)
- data limitations.

Estimation of allowable source concentrations requires assessment of exposure or calculated dose, which includes assessment of a range of factors such as allowable daily intake, ingestion/inhalation rate, fraction ingested, body weight, frequency, duration and period of exposure.

In the UK, current models include CLEA and SNIFFER, both of which are derived from initial research by the late Prof. Colin Ferguson at Nottingham University. (NB: CLEA 2002 has been withdrawn and CLEA UK, Beta version 1.0, dated 2005, is only available for user evaluation and comment, with a final version of the software expected late 2007.)

The models have limitations, particularly with respect to certain exposure pathways and generic land uses. Their applicability should be assessed against the particular conceptual model and contaminant behaviour before use and adoption of assessment values so derived. Other non-UK models are available which include different exposure scenarios such as BP RISC,[58] RiscHuman[59] and RBCA.[60] The applicability of these to the UK and the variations from the UK standard exposures must be carefully assessed before use and, if appropriate, agreed with the regulators.

Groundwater assessment uses fate and transport modelling, typically using multi-tier two-dimensional models such as ConSim or R&D P20 to determine contaminant concentrations either at the receptor due to a known source or to determine allowable concentrations within the source (i.e. clean-up targets) based on acceptable limits at the receptor, respectively. ConSim also has the attribute of being a probabilistic model (in its current version) while P20 is deterministic. However, in using these models, it is critical that the conceptual model for the site correlates with or can be adapted to the model used within the assessment package. For the most complex situations, it is also possible to use three-dimensional modelling using such analytical packages as Modflow.[61] The use of such models normally requires highly specialist hydrogeologists and, as with all computer analysis, its use requires appreciation of whether the results from such assessments are meaningful and known input parameter variations allow 'strict' interpretation.

Gas risk assessment models have been developed such as CIRIA report 152, which uses a probability tree approach to assessing gas risk in buildings. GasSim[62] is a fate and transport model which uses information on waste composition and quantity, landfill engineering, and landfill gas management techniques to enable assessment of the best combination of gas control measures for a particular design and rate of filling.

Analogous procedures can be developed to cover other types of hazards (e.g. calculation of the concentration of an explosive gas in air, taking into account source concentrations, migration characteristics, and circumstances of exposure). Exposure and toxicity assessments are typically subject to many uncertainties due to information gaps in the effects and toxicity assessments. While every effort should be made to reduce uncertainties, for example by collecting more detailed site investigation data, it may be necessary to make assumptions in order to complete an assessment. A common approach is to apply a 'worst case' scenario so that sufficient safety margins are built into the assessment, but this can lead to over-conservative and unrealistic modelling. Some assumptions may be present in the default values built into models. In all cases it is essential to identify and record all uncertainties and assumptions used in the assessment. It is also important to ensure that the assumptions used will stand up to scrutiny and that sensitivity of the risk estimation to the different assumptions involved is properly appreciated.

3.3.4 Risk evaluation

Risk evaluation involves making judgements about the acceptability of risk estimates, having regard to available guidance and taking into account any uncertainties associated with the process. It is important to remember that initial judgements about the acceptability of a risk may be modified at a later date when the costs and feasibility of taking remedial action have been more fully evaluated (see Chapters 4–7). For example, a risk may be considered unacceptable (even when judged to be low) if there are serious consequences (e.g. an explosion leading to human fatalities). A high risk (e.g. death of a proportion of young landscape plants) may be tolerated if the cost and practical problems of removing the source of the risk (moderately high concentrations of phytotoxic metals) are more onerous than those associated with rectifying the damage (e.g. periodic replacement of stock) should it occur. Different parties may also have different views on what constitutes an 'unacceptable' risk.

Although there are published guidance values for soil, there is currently significant debate at national level whether these are appropriate limits to set for determining contaminated land as compared to their use in the design process. It is accepted that exceedance of an SGV represents a risk, but it is argued that this risk is not necessarily *significant* in accordance with the definition in Part 2A (see section 1.1). Therefore, the risk evaluation process must be transparent and clearly documented, as well as based on a sound scientific approach, if it is to be defensible, especially if the assessment is being used in the context of Part 2A.

An important task in risk evaluation is testing the sensitivity of the outcome of the assessment to changes in the assumptions used. This is particularly important in marginal cases where relatively small adjustments to the assumptions may have a significant effect on the risk estimate and major implications for the type and cost of any remedial action.

3.3.5 Risk mitigation

It is important to recognise that any residual contamination equates to an acceptable level of risk generally following assessment against an appropriate set of criteria, be it published limits or following use of quantitative risk assessment modelling. If the source concentrations exceed acceptable criteria and the levels of risk are deemed unacceptable, then remedial action to reduce source levels, break pathways or remove receptors will be required. It is essential that any approach be agreed in close consultation with the regulatory authorities, and other third parties (e.g. funders, insurers, etc.) as appropriate.

3.4 Reporting

Factual reporting is often seen as a matter of routine and one whose quality is addressed by a string of box ticking and quality audit signatures. However, it cannot be emphasised enough that it is the data on which all assumptions are made, conclusions drawn and recommendations established for any site. Therefore 'testing' the reliability of the reported data is paramount.

Questions such as the following should be asked:

- Is it reasonable to expect this heavy metal at this concentration, can it be achieved outside mineral bearing horizons?
- Is there an alternative reason for high methane concentrations more connected to the monitoring installation and groundwater water levels than a soil-borne source?
- Are these results expected and if not is there a rationale for their potential validity?

Reporting should aim to give the client and regulators a clear understanding of the type and source of contamination as far as is possible and demonstrate that the site conceptual model is valid or has been modified according to investigation data.

Further, it is unacceptable to leave an interpretative report as just a list of perceived problems or risks often accompanied by extremely detailed discussions of testing protocols and legislatory background. The instigator of the report almost certainly perceives some element of risk already. What is required by both regulators and clients is a clear understanding of the inter-relationship of the ground conditions and contamination, coupled with an assessment of the risk to current good practice and a clear set of practically feasible mitigation measures that can be validated at the end of the works. This is the vital element of interpretation, albeit based on information that may be less comprehensive, lacking long-term monitoring or other factors that commercial circumstances place upon professionals.

The site investigation report is perceived as a marker point – even an end point in the process for some members of the professional team – and it may involve changes in personnel or companies. In some instances, clients simply do not appreciate the difference between a factual report and an interpretative one and are perplexed by Remediation Method Statements and Validation Reports. Each type of report has an approach and a format that individual organisations favour, however both the AGS[63] and BS 5930 identify typical contents of interpretative reporting.

The following should be seen as amplification of these frameworks and an interpretative report should include:

(*a*) A description of the decision-making framework including investigation objectives. In particular, any constraints, such as a lack of time, inadequate financial resources, practical difficulties (e.g. restricted access to parts of the site) that may have applied to any phase of investigation and risk assessment.
(*b*) An appropriately detailed desk study and identified site conceptual model if a separately issued study does not already exist.

(*c*) A factual account of all the work carried out, supported, as appropriate, with graphical material in the form of maps, photographs, sampling and analytical procedures, etc.
(*d*) The findings of the investigation including field observations, analytical results, monitoring data, etc.
(*e*) The statistically based, assessment of contaminants.
(*f*) Appropriate risk assessment (qualitative or quantitative) with an analysis of uncertainties, and details of any assumptions, safety factors, etc. used.
(*g*) Summary of remediation recommendations including analysis of any technical or practical constraints that are likely to affect the type of remedial action that can be taken at the site.

It is important that there is recognised consistency with geotechnical findings and recommendations so that a coherent and feasible set of solutions is presented and 'vacuum' recommendations are not produced.

The Environment Agency has published eight checklists[64] for activities and reporting at various stages of investigation, risk assessment, remediation and validation. These may be used in some circumstances as a guide to report formats.

Part II.   Remediation

4. Remediation as part of risk management

4.1 Introduction

Risk reduction through validated remedial action is the final element in managing the risks associated with contaminated land (see Chapters 1–3). It comes into play when, on the basis of a risk assessment, it is decided that the site poses unacceptable risks to specified targets and action should be taken to reduce or control the risks to an acceptable level.

Typically, a number of remediation options will be available to the assessor, but only a few (or combinations thereof) will offer an overall balance between technical effectiveness, practicality and cost. The selection will also be influenced by sustainability issues and, in certain circumstances, forthcoming changes to legislation, such as the ongoing staged implementation of the European Landfill Directive. There may also be conflicts between acceptable technical solutions and public perception of what is acceptable.

The purpose of risk reduction is therefore to:

- specify acceptable levels of risk reduction/control
- identify a remediation strategy that will meet risk reduction/control and other objectives, and satisfy the regulators
- design and implement the strategy
- ensure, through a programme of monitoring and validation, that remedial objectives have been met.

Remedy selection, design and implementation normally follow on from site investigation and assessment, and the identification of acceptability criteria or site specific target levels (SSTLs) (see Chapter 3). However, there is overlap between the various stages and there may be a need to carry out supplementary investigation or other studies before a detailed design can be finalised.

The following chapters are concerned principally with the use of technical measures for controlling/reducing the risks associated with contamination on land. However, it is possible to control risks using administrative means as below.

(*a*) Adopting a 'less sensitive' use of the land (e.g. substituting commercial/industrial for residential/horticultural use: formal restrictions on the use of the land may have to be made).

Figure 4.1 Cut-off and capping construction

(*b*) Restricting access to the site.
(*c*) Altering the form or layout of a development (e.g. to avoid areas of severe contamination), subject to planning constraints.

Administrative measures are typically not applicable where the contaminant is mobile and migrating off-site and, as a result, is threatening sensitive surface or groundwater bodies or other targets. However, in specific circumstances, adoption of Monitored Natural Attenuation as a solution to groundwater contamination can provide an administrative solution.

Administrative measures may be effective where hazards, pathways and receptors are specifically related to the use of the land (e.g. direct access by children to contaminants at the surface, direct contact by service pipes to contaminants in the ground). The use of administrative measures alone also means that there is no material improvement in the condition of the land: the same situation will have to be confronted again in the future should the condition or use of the site change.

Where administrative means are used to control or reduce risks, it may be necessary to include control measures in site management documents or covenants on deeds for properties, which have to be disclosed in the event of sale. For example, restricting the size and depths of water features in gardens or conservatory construction where capping or membranes might be punctured.

In addition, it may be possible to obtain project specific insurances to guard against unforeseen costs associated with remediation being required at a later date.

4.2 Remedy selection

Remediation methods selected for use must be:

- applicable to the contaminant and media (e.g. soil, sediment, construction debris, surface/groundwater, etc.) to be treated;

Options **1.** Do nothing
2. Dig and dump
3. Groundwater treatment
4. Stabilisation
5. Containment

Weights Best estimate of relative importance of each criterion
Cost = 25%
Time = 25%
Space requirements = 25%
Impact on estuary = 25%

Costs, including monetary, time, space etc., are assigned a negative value. (Benefits are positive)

Criteria	Cost (£)		Time (weeks)		Footprint (m^2)		Impact		C_j
Weight (*W*)	0.25		0.25		0.25		0.25		
Option	*c*	$W.c/\lvert c_{max}\rvert$	*c*	$W.c/\lvert c_{max}\rvert$	*c*	$W.c/\lvert c_{max}\rvert$	*c*	$W.c/\lvert c_{max}\rvert$	
1.	–1	0.00	–1	0.00	–1	0.00	–1	–0.25	–0.25
2.	-3.00×10^6	–0.25	–5	–0.01	–1000	–0.25	1	0.25	–0.26
3.	-1.00×10^6	–0.08	–100	–0.25	–450	–0.11	4	1.00	0.55
4.	-1.50×10^6	–0.13	–6	–0.02	–1000	–0.25	5.5	1.38	0.99
5.	–100 000	–0.01	–1	0.00	–300	–0.08	8	2.00	1.91
	$\lvert c_{max}\rvert$ = 3 000 000		$\lvert c_{max}\rvert$ = 100		$\lvert c_{max}\rvert$ = 1000		$\lvert c_{max}\rvert$ = 10		

Notes
C_j – overall criterion function. Preferred alternative is that with highest *C* i.e. most positive C_j value $= \sum W.c/\lvert c_{max}\rvert$ where *W* = weighting, *c* = cost or benefit, $\lvert c_{max}\rvert$ = maximum cost of benefit (absolute value)

Figure 4.2 Example evaluation matrix

– effective in achieving specified acceptability criteria or SSTLs;
– feasible in the sense that they can be put into practical effect;
– acceptable to all relevant parties including the regulators;
– economic;
– sustainable.

In practice, each remediation method has advantages and limitations that may constrain its use on a site-specific methods basis. The purpose of remedy selection is to identify and then evaluate remedial methods (or combinations of methods) which may be suitable for use on a particular site, with the aim of identifying that remedial strategy (the preferred remedial strategy) best able to satisfy site-specific remedial objectives and overcome constraints.

It can be useful to provide an options study at the selection stage, and may be requested by the regulators, especially where sites are particularly sensitive. Quantified options studies that provide a weighted matrix approach to the influencing factors can be used effectively to determine the optimum solution for a particular site, with weighting factors set according to the relative importance of the various influences (see Figure 4.2). Sensitivity studies are also useful, in assessing which aspects have greatest influence on the selection of methodology.

4.3 Design and implementation

Design and implementation is that part of risk reduction that transforms a remediation strategy from a conceptual state into practical action at the site.

Largely, the same principles of design and implementation employed in conventional civil engineering/construction applications can be applied to the remediation of contaminated land: in some cases, the same practical techniques can be employed.

However, the presence of hazardous substances on contaminated sites, and the relative lack of well established guidance in this field (e.g. Codes of Practice, standard contract terms and conditions, technical specifications, etc.) means that conventional approaches and tools may not be directly applicable. In some cases (e.g. the application of a chemical or biological treatment process) they may be of very limited value. Other elements of remediation design and implementation, such as the use of treatability testing or pilot trials in the early stages of selection and design, are largely unfamiliar in a conventional civil engineering context and special provision may have to be made to incorporate these elements into the design and implementation process.

The process of design and implementation of a remediation strategy can involve a significant period of time. Time must always be allowed for the consultation period with the regulators that can be in the order of a month at each stage of a query. Although most officers will try to respond in an appropriate timescale for the circumstances, they are not obliged to do so. It is important to recognise this, particularly in the context of a redevelopment project where there may be pressures to reduce or minimise the scheme programme.

Key aspects of design and implementation in land remediation applications are listed below.

For the design process:

- identifying objectives and constraints;
- planning, design and specification;
- procurement options;
- contract options;
- health and safety considerations (CDM).

For implementation:

- project supervision;
- communication;
- monitoring and validation;
- post-treatment management (where appropriate);
- documentation and reporting (validation).

These aspects are important because they help to ensure not only that remedial action will achieve its objectives, but also that comprehensive evidence is collected to demonstrate that remediation objectives have been met.

In the housing sector of the construction industry, lack of evidence can have severe consequences. Under current UK guidance for house builders, failure to demonstrate adequate remediation can result in non-compliance with planning conditions and non-acceptance by the planning authority. 'Red carding' by the NHBC results in no warranty being available for the property and, under the Council for Mortgage Lenders initiative, no funding being available from lenders. Therefore, a lack of records can result in property sales being jeopardised.

5. Remediation options

5.1 Classification and terminology

There is no universally accepted classification of, or terminology for, remediation methods. For the purposes of this Guide, the approach used in the CIRIA guidance document on the Remedial Treatment of Contaminated Land[2] has been adopted and modified.

Currently available methods can be classified into three broad groups (see Figure 5.1):

(*a*) Civil engineering-based methods: these employ conventional civil engineering techniques to remove or contain contamination sources, or to block the pathways by which contaminants reach receptors.
(*b*) Process-based methods: these use specific physical and chemical processes to remove, immobilise, destroy, or modify contaminants.
(*c*) Natural remediation: allowing existing or introducing natural processes to effect reduction.

The groups are not mutually exclusive and there are remediation methods used under certain circumstances, such as the use of permeable reactive barriers, which fall into more than one. However, their classification provides a useful basis for considering remediation methods available for treating contaminated soils, sediments and water. A remediation strategy may incorporate methods from each group, or use a number of different methods on an integrated basis, to achieve specified remedial objectives.

Other important terms that may be encountered in a remediation context are:

Ex-situ remediation: where treatment is applied following the excavation (in the case of solids) or extraction (in the case of liquids and gases) of contaminated material.

In-situ remediation: where treatment is applied to contaminated media without prior removal from the ground.

On-site and off-site remediation: relate to the location of treatment. All in-situ treatment is carried out on the site undergoing remediation, but ex-situ treatment may be carried out on site (for example using mobile treatment plant) or at centralised/merchant facilities off site (e.g. hazardous waste incineration, physical–chemical waste treatment, etc.).

Treatment: used in its normal sense to imply some material improvement in the condition of a site or waste materials. It is not intended as a comment on the quality or permanence of a particular method.

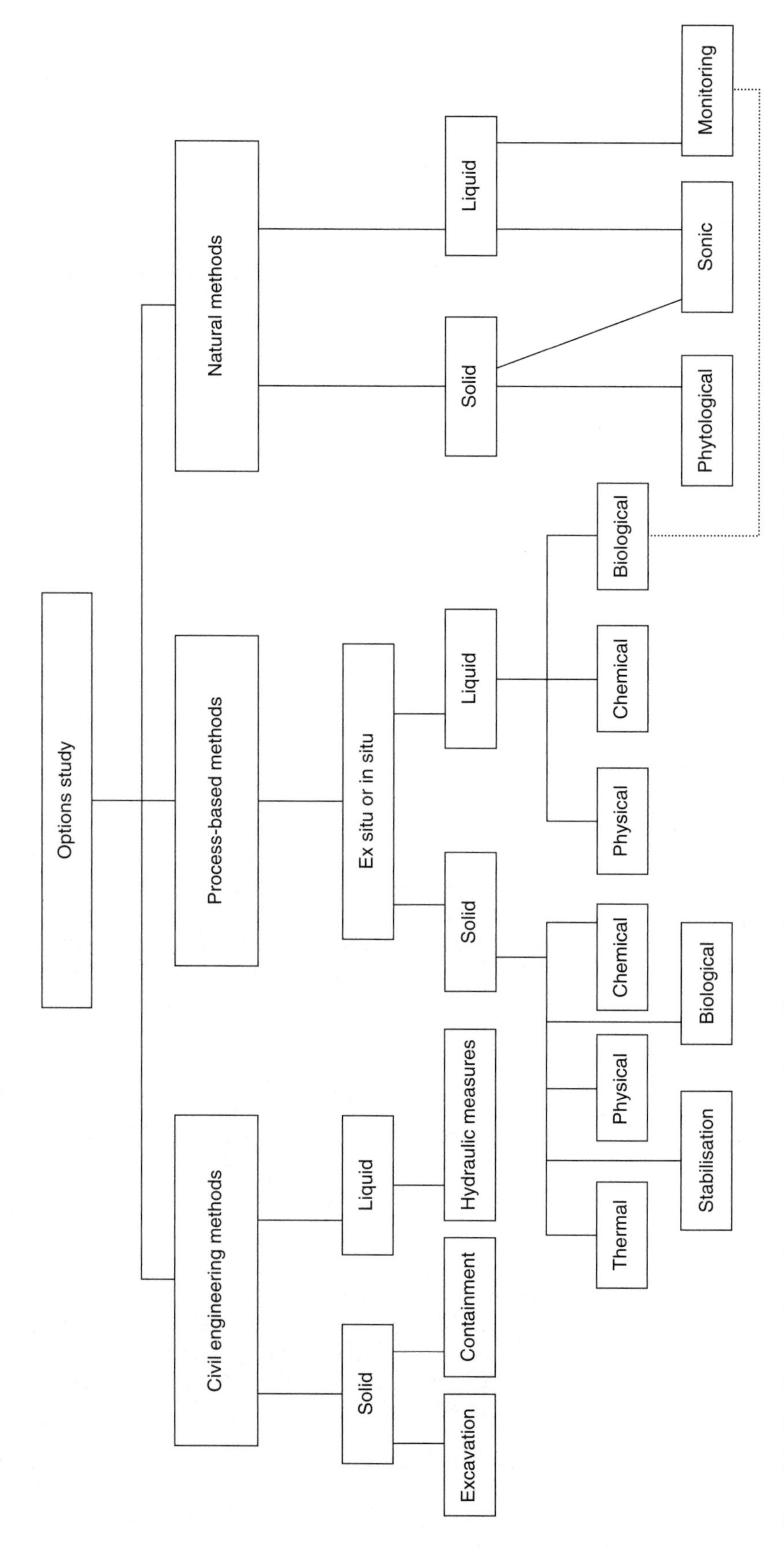

Figure 5.1 Classification of remediation methods (revised from The Remedial Treatment of Contaminated Land, CIRIASP101-112, 1994)

(Note: in the context of waste disposal, treatment of hazardous waste is required to pass the 'three point test', i.e. a physical, thermal, chemical or biological process (including sorting) which changes the characteristic of the waste so as to reduce volumes, reduce its hazardous nature or improve handling and enhance recovery.)

5.2 Civil engineering-based methods

Civil engineering-based methods can be classified into three main groups:

(*a*) **removal** (excavation) of contaminated solid material.
(*b*) **physical containment** (of the contaminated ground) using covers and in-ground barriers.
(*c*) **hydraulic controls**, used in support of (*a*) and (*b*) above; as the principal means of control; or specifically for the treatment of contaminated surface or groundwater.

Some soil treatment methodologies, such as soil mixing/stabilisation, utilise traditional civil engineering plant, but their success depends on the chemical reactions, so these are included under the process-based methods.

In general, civil engineering-based methods are relatively insensitive to variations in the concentrations and types of contaminants present, or the types of contaminated media being handled. In this respect, they are of potentially wide applicability. They are also well established, familiar to both designers and contractors, and use readily available plant and equipment. However, they suffer from a number of limitations: excavation may pose health and environmental impacts; containment systems do not materially reduce the volume or the hazardous properties of contaminated material; they have a finite life and their effectiveness is thought to decrease over time.

Removal of contaminated materials to landfill has been recognised as generally unsustainable and the implementation of the Landfill Directive was intended to

Figure 5.2 Jet grouting

reduce the amounts of waste disposed of in this way. This has resulted in stricter controls on what can be taken to landfill and redefined what constitutes hazardous waste. Associated new licensing regulations have reduced significantly the number of landfills licensed to accept the most contaminated soils. This is discussed further in Appendix F.

More detailed information on the capabilities and limitations of individual civil engineering-based methods of treatment are provided in Appendix C.

5.3 Process-based methods

Process-based methods can be classified into five main generic types:

(*a*) **thermal treatment**: using heat to remove, stabilise or destroy contaminants
(*b*) **physical treatment**: using physical processes, such as mixing or exploiting physical attributes, to separate contaminants from host media, or different fractions of contaminated media (generally uses traditional civil engineering-based processes to facilitate the treatment)
(*c*) **chemical treatment**: using chemical reactions to remove, destroy or modify contaminants; includes stabilisation and solidification, where contaminants are chemically stabilised and/or immobilised to reduce their availability to receptors
(*d*) **biological treatment**: using natural metabolic pathways of micro-organisms and other biological agents to remove, destroy or modify contaminants (see also natural remediation)
(*e*) **stabilisation/solidification**: in which contaminants are chemically stabilised and/or immobilised to reduce their availability to targets.

Compared to civil-engineering methods, process-based methods of treatment have much more specific capabilities and requirements. As a result, they tend to be restricted to a more limited range of contaminants and media. However, many have the advantage of reducing the volume or concentration of hazardous substances in affected media and, if they also destroy contaminants, may provide a more 'permanent' solution to the contamination.

Generic processes can be applied in either an ex-situ or in-situ mode. In-situ applications avoid the cost and potential above-ground environmental impacts associated with excavation/extraction. However, at the present time, there is generally less practical experience in the application of in-situ methods compared with their ex-situ counterparts. It may also be more difficult in practice to predict (or demonstrate) the outcome of an in-situ application, or to optimise/control the process once in operation.

More detailed information on the capabilities and limitations of individual process-based methods of treatment is provided in Appendix C.

5.4 Natural remediation

Natural remediation includes monitoring and measuring those processes that occur naturally in the environment, and in some cases encouraging and enhancing those processes to take place. Strictly, this is a form of bioremediation, although remediation itself is as a treatment since the methodology requires the soil to be processed in some way to achieve a uniform and consistent effect.

These methods are generally attempted when the success of more 'traditional' methods is likely to be limited, unsustainable, logistically almost impossible or economically unviable. However, they should not be viewed as a 'cheap fix' or 'do-nothing' option as they involve a considerable commitment to initial feasibility studies and, subsequently, of time and money to achieve and record any results. The validation process can also be long term, complicated and difficult, especially if regulations change during the lifetime of the remediation.

The following list identifies some commonly used 'natural' methods and other more experimental natural processes with a potential to become natural remediation technologies:

(*a*) Degradation (by do-nothing).
(*b*) Monitored attenuation, i.e. the term applies typically to measurement and monitoring of degradation of contamination in groundwater.
(*c*) Phyto-remediation – use of plants to remove metallic contaminants from soils.
(*d*) Fungal innoculation – use of natural fungi (typically white rot fungi) to degrade recalcitrant organopollutants such as polyaromatic hydrocarbons (PAH), organo pesticides, munitions, bleaches, dyes, polychlorinated biphenyls (PCBs). Commercial application of such methodologies has not yet been established.
(*e*) Sono-chemistry – use of ultrasonic energy to promote chemical and physical reactions. By sonicating a liquid, powder or compound chemical and physical reactions can be speeded up or can be made to form new compounds. The immense temperatures, pressures, the quick heating, and cooling cycles generated on a microscopic scale by the ultrasonics, enable high-energy chemistry. However, the effectiveness on anything other than lab scale trials has yet to be demonstrated.

6. Remedy selection

6.1 The selection process

The prime requirement of remediation is to break the pollutant linkage between identified sources and receptors. There may be a number of linkages involved. In some cases, a strategy based on a single remediation method may be sufficient to address all of the risks presented by the site. In others, a range of different methods may have to be combined to provide an integrated remedial strategy capable of dealing with different types of contaminants, or different parts of the site. Although the acceptability criteria or SSTLs should have priority in determining the type of remedial action taken, it will usually be necessary to meet other technical objectives (e.g. on the engineering properties of the site) or environmental/sustainability criteria (i.e. waste reduction, limitation of traffic movement, dust and noise limitations) and overcome practical and cost constraints. The main purpose of selection is therefore to:

(*a*) identify those remediation methods most likely to be applicable, effective and feasible on a site-area or media-specific basis;
(*b*) develop a range of potentially applicable, effective and feasible strategies based on a short-list of favourable methods; and
(*c*) evaluate individual strategies (which might comprise one or a combination of specific methods) to determine which is most likely to offer the best balance between technical effectiveness, practicality and cost.

Selection is a staged process of identification, evaluation and screening (see Figure 6.1) by which some remedial methods (or strategies) are retained for further consideration and others are rejected. The use of selection criteria (see Selection criteria and procedures, below) can help to keep the process both manageable and objective.

A structured framework, such as detailed in CLR 11, encourages thorough consideration of available options (and their practical and cost implications), and helps to avoid the development of inappropriate strategies based on too little data and an inadequate assessment of alternatives. It also helps the designer to anticipate the potential benefits and limitations of different types of approach that otherwise might not be readily apparent. By requiring the designer to present the rationale for acceptance/rejection decisions, the selection process also makes decision-making more transparent and accessible to third parties. This is important when discussing a proposed course of action with the regulatory authorities and the local community and, in the future, when a remediated site is sold or its ownership transferred.

The scope of the selection process and the way in which decisions about remediation are made in practice will vary to an extent depending on the exact circumstances of the

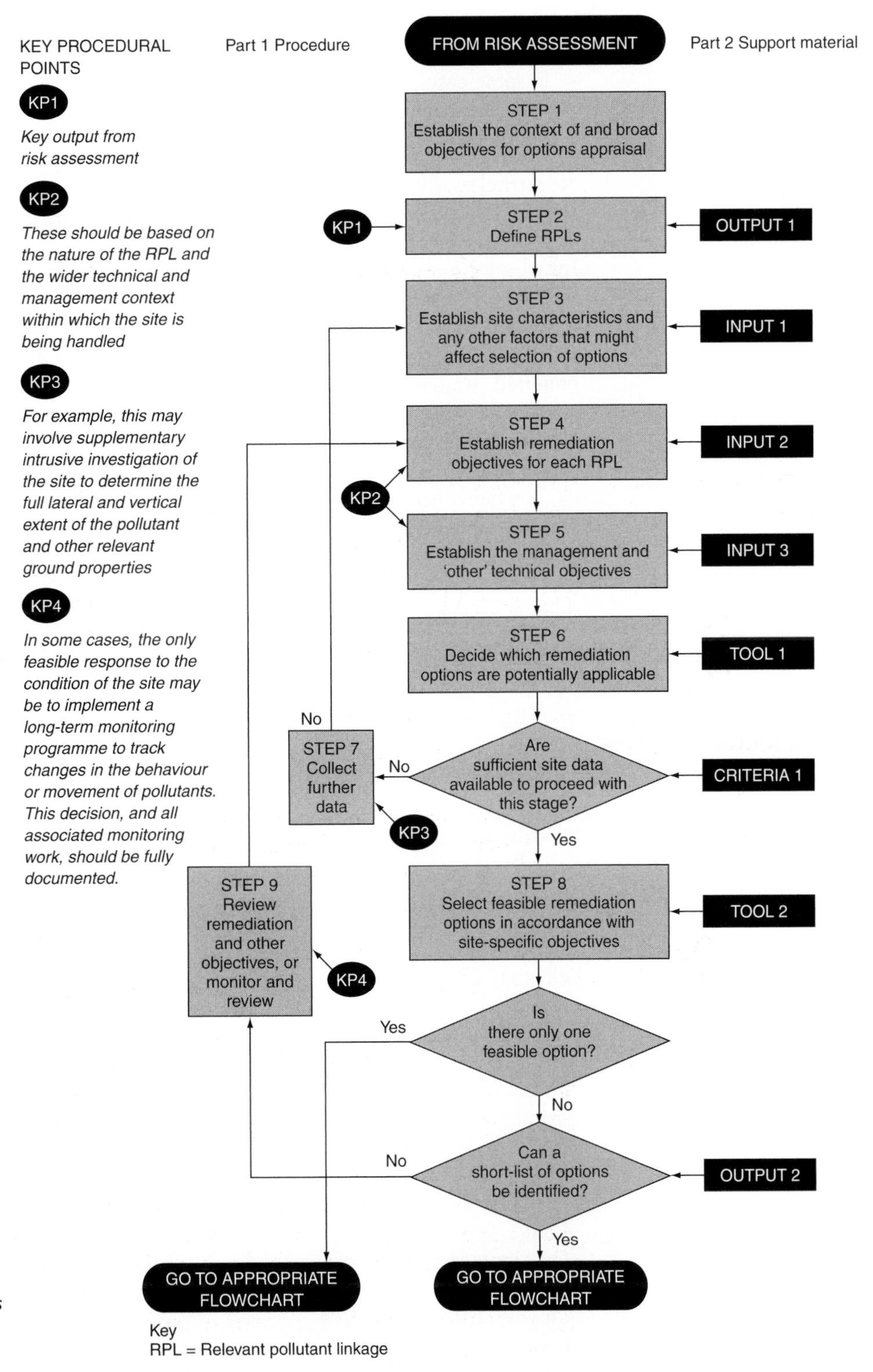

Figure 6.1 The selection process – example flowchart from EA publication CLR11[14]

site, the experience of the designer and the stage in the selection process. For example:

– preliminary views on potentially useful remediation methods may already have been formed during the detailed investigation and assessment of the site

– it may be apparent at a very early stage that only a limited number of remediation options exist
– in some cases, it may be necessary to review remediation objectives in the light of more information on site characteristics, and the capabilities and limitations of the various remedial methods.

Nevertheless, for the reasons outlined above, it is important to present a full justification of all acceptance/rejection decisions made during the selection process.

Initial selection decisions are typically made on the basis of experience and published information readily available to the designer. Initial selection criteria (see below) can help in deciding which remedial methods are most likely to be suitable, given the specific circumstances of the site and its current or planned use. Selection of the preferred strategy should be based on a detailed analysis of a limited range of alternative remedial strategies. The use of final selection criteria and formal ranking systems may be beneficial during the detailed assessment stage. In some cases (particularly where process-based remediation methods are under consideration) it may be necessary to carry out bench or pilot-scale treatability studies (see 6.2.4 Role of treatability studies) and/or supplementary site investigation to complete the detailed assessment. It is also essential for designers to involve specialist contractors where appropriate, in assessing the economic, technical and logistical feasibility of a particular process. For example, there is no value in pursuing soil washing as a solution if there is insufficient space to accommodate the plant, plus the various stockpiles of untreated, treated and waste material and allow for material and plant movements.

6.2 Selection criteria and procedures

6.2.1 Initial selection and evaluation

Initially, remediation methods should be selected based on their:

– applicability (to contaminants and media)
– likely effectiveness (in meeting acceptability criteria and other technical objectives)
– feasibility.

Selection criteria (see Box 6.1) can be used to help in the initial screening and evaluation of methods. Further information on initial selection criteria and their role in the selection process is summarised in Appendix E.

Box 6.1 Initial selection criteria

Applicability	Time constraints
Effectiveness	Planning and management needs
Limitations	Health and safety needs
Cost	Potential for integration
Development status/track record	Environmental impacts/benefits
Availability	Monitoring
Operational requirements	Validation
Information requirements	Post-treatment management
Licensing and legal issues	

The output from the first stage of selection should be a short-list of potentially viable remediation methods classified according to the media/zones of the site to be treated, approximate volumes of material involved, and remediation objectives and criteria to be met.

6.2.2 Development of remedial strategies

Once a range of remediation methods for treating particular contaminants or parts of the site have been identified, those methods should be combined to develop a flexible range of strategies for treating the site as a whole. A number of alternatives should be developed in the first instance so that there is some scope for choice during final selection, when the technical sufficiency, feasibility and costs of alternative schemes are evaluated in detail. Unless the circumstances of the site dictate otherwise, a range of alternative strategies, from minimal action (perhaps at the expense of long-term security or limitations on the use of the site) to comprehensive action (perhaps at high initial costs) should be developed.

Combining remediation methods for operation on an integrated basis may have technical, practical and economic implications. Checks should be made to ensure that remediation strategies are likely to remain applicable, effective and practical. It should be noted that savings in remediation costs can be negated by disproportionate mobilisation costs where multiple processes are adopted.

This stage of the evaluation also provides an opportunity to check:

- that remediation objectives and criteria remain valid and likely to be achievable in practice
- whether more detailed information, for example on site conditions or the anticipated performance of remediation methods, is needed in order to complete the assessment.

In the first instance, further desk-based research should be carried out to resolve any information gaps. It is important to bear in mind, however, that different remediation methods have quite specific information requirements and it may be necessary at this stage to carry out additional (i.e. supplementary) site investigation, or treatability studies, to provide the required data.

The output from this stage of selection should be a limited number of potentially applicable, effective and feasible remediation strategies that can then be taken forward for detailed evaluation.

6.2.3 Selection of preferred remedy

In this stage of selection, alternative remediation strategies are analysed in detail to determine their respective strengths and weaknesses. Sensitivity analysis may also be carried out to determine the optimum choice in the event of predicted possible or likely changes in circumstances, e.g. significant cost increase in disposal to landfill. Final selection criteria, addressing both long- and short-term objectives and operational issues as below (see Box 6.2), and formal ranking systems (see Table 6.1) may help in the assessment process.

Box 6.2 Examples of final selection criteria

Long-term criteria	Short-term criteria
– Legal/regulatory compliance – Long-term effectiveness – Reduction in toxicity, mobility and volume – Acceptable track record of use – Acceptability to local community	– Acceptable operational requirements – Minimal short-term health and safety implications – Minimal short-term environmental impacts

Table 6.1 Example of formal ranking procedure for alternative remedial strategies

Criterion	Rank*	Weighting for each strategy†			Overall score, for each strategy		
		A	B	C	A	B	C
Legal/regulatory compliance	10	3	3	2	30	30	20
Long-term effectiveness	8	3	1	1	24	16	8
Reduction in hazard	6	3	2	1	18	12	6
Track record	4	3	3	3	12	12	12
Acceptability to local community	2	2	3	1	4	6	2
Operational requirements	2	1	2	3	2	4	6
Short-term health and safety implications	2	1	2	3	2	4	6
Short-term environmental impacts	2	2	2	1	4	4	2
Overall score					96	88	62

Note: values in table are for illustrative purposes only.
* Ranks and weighting factors should reflect site-specific priorities.
† High weighting reflects favourable attributes (e.g. good long-term effectiveness, insignificant operational requirements, etc.).

The outcome of ranking, together with information on costs and the ability of each strategy to meet overall technical and administrative objectives, can then be used as a basis for selecting the preferred remedial strategy.

However, it should be noted that this process is open to bias, depending on how the weightings are set, also depending on the sensitivity of the matrix can be mislead. However, the approach provides a framework provoking thought and reasoned evaluation and is viewed favourably by the regulators as a method of demonstrating the decision process.

6.2.4 Role of treatability studies

Treatability studies are carried out to:

- provide site-specific information on the likely technical performance of a particular, process-based remediation method or strategy
- reduce technical and financial uncertainties associated with particular remediation strategies.

The need for treatability data may be identified during the initial screening of methods, or during the final assessment of alternative strategies. Because of the costs involved, treatability testing on any significant scale is only likely to be feasible when the designer is sure that a particular method has a reasonably good chance of being applied (i.e. once a method has reached the final assessment stage). However, it is important to note that the absence of treatability data at an early stage can result in an inappropriate method progressing a long way down the selection pathway before being eliminated. Since the need for treatability studies introduces a further stage into the process of remediation strategy selection and implementation, more time is required before the remediation work can be completed. Treatability studies can sometimes take a significant period of time and this may itself become a factor in the selection process.

Treatability tests should always be carried out where a method does not have a documented track record of use in the field, or where insufficient data are available to predict the likely outcome of treatment even where the contaminated matrix has been well defined.

Box 6.3 Issues to be addressed when developing treatability test places

- Background information and rationale
- Site characteristics
- Properties of test material
- Test objectives
- Procedures, equipment and materials
- Analytical methods
- QA/QC plan
- Data management
- Data analysis and interpretation
- Health and safety
- Waste management
- Contingency planning

Tests can be conducted at either bench or pilot-scale. It is important to document fully both the test plan (e.g. aims, methods, means of interpretation, etc.) and the outcome (results, interpretation, conclusions, recommendations, etc.). Issues to be addressed when developing a treatability test plan are summarised in Box 6.3. Typically, treatability studies are carried out by specialist contractors. Further information on the design and application of treatability studies for the remediation of contaminated land is given by CIRIA SP 164.

Note: parameters apply to bench and pilot-scale testing: additional requirements for pilot-scale work are:

- pilot plant installation and start-up
- plant operation and maintenance
- operating conditions to be tested
- sampling plan (for operational trials).

7. Design and implementation

7.1 Introduction

Design and implementation transform the preferred remediation strategy from its conceptual state into practical action in the field.

Site investigation, assessment and remedy selection should have been sufficient to:

- establish the overall goal of remedial action (i.e. to reduce/control the risks associated with contaminated land to acceptable levels)
- identify specific objectives (e.g. acceptability criteria/SSTLs) for remedial action
- identify specific constraints to remediation activities
- provide some idea (outline design) of the works needed to achieve stated objectives.

The next stage is to develop these outline requirements into a more detailed set of objectives and plans that can be used to implement the remediation strategy and demonstrate that remediation objectives have been met.

7.2 Planning and design

7.2.1 Project planning

The main purpose of project planning is to establish:

- what activities should take place to achieve the objectives of the project
- the relationship between the activities
- what resources (including time) should be made available to enable the activities to take place.

The advantages of project planning are that it:

- creates a more accurate picture of what should happen as a project progresses
- allows better anticipation of what needs to happen next
- allows critical points in a project to be identified
- permits more reliable estimates of resource requirements to be made
- allows better anticipation of, and planning for, bottlenecks in a project
- enhances cooperation and communication
- helps to build commitment between project participants
- assists in completion of the project according to the required standards, on time and within budget.

The results can be usefully presented in pictorial form, for example as bar or Gantt charts (as shown in Figure 7.1), or critical path charts. It is essential that sufficient time and resources are made available for project planning, and that it addresses both the management and technical aspects of the remediation works.

7.2.2 Technical specification

The technical output of project planning is the detailed specification and programme. Clearly, these will vary depending on the circumstances of the site, the type of remedial action being proposed and the management objectives (e.g. timescales, costs, procurement preferences, etc.) agreed for the project. Many of the technical aspects should have been addressed during investigation, assessment and remedy selection. The detailed technical specification should therefore build on the information already assembled, taking care to address the main aspects listed in Box 7.1.

Box 7.1 Aspects to be addressed when developing the detailed specification

Remediation objectives and constraints (e.g. by site area, contaminant type)
Other technical objectives (e.g. engineering properties)
Environmental objectives (e.g. sustainability, habitat protection, environmental management systems)
Proposed remediation methods (mobilisation, procedures, equipment, supply services, layout, phasing, materials handling and waste disposal)
Legal approvals, licenses and conditions (see below)
Duration, phasing and integration (within remediation strategy itself and externally, e.g. with construction, landscaping, etc.)
Public health protection measures (e.g. containment measures noise, vehicle access and movements, road cleaning)
Safety (e.g. liaison with planning supervisor/site coordinator, method statements, health and safety plan, site inductions, tool-box talks/awareness training, signage)
Occupational hygiene (equipment and procedures)
Site preparation (e.g. access, security, site services, site storage, laboratory support, decontamination facilities, etc.)
Monitoring requirements (for legal compliance, process control and optimisation, demonstrating long-term performance)
Validation requirements (e.g. post-processing sampling and analysis, pre- and post-remediation surveys, photographs, quality of excavated area, quality of materials used in remedial work)
Public relations (e.g. information needs and procedures, contact points, public meetings, etc.)
Post-treatment management (e.g. monitoring, administrative controls)

The technical specification should describe the scope of the proposed remediation work and the standards (e.g. of remediation, workmanship, materials, etc.) to be achieved. It may be used to obtain contracting services through formal processes of tendering or negotiation.

Standard specifications may be available, or applicable with some adjustments, for some types of engineering-based remediation strategies. However, in most cases, the specification will be bespoke to the project and may have to be developed from first principles (see Section 7.3 'Procurement').

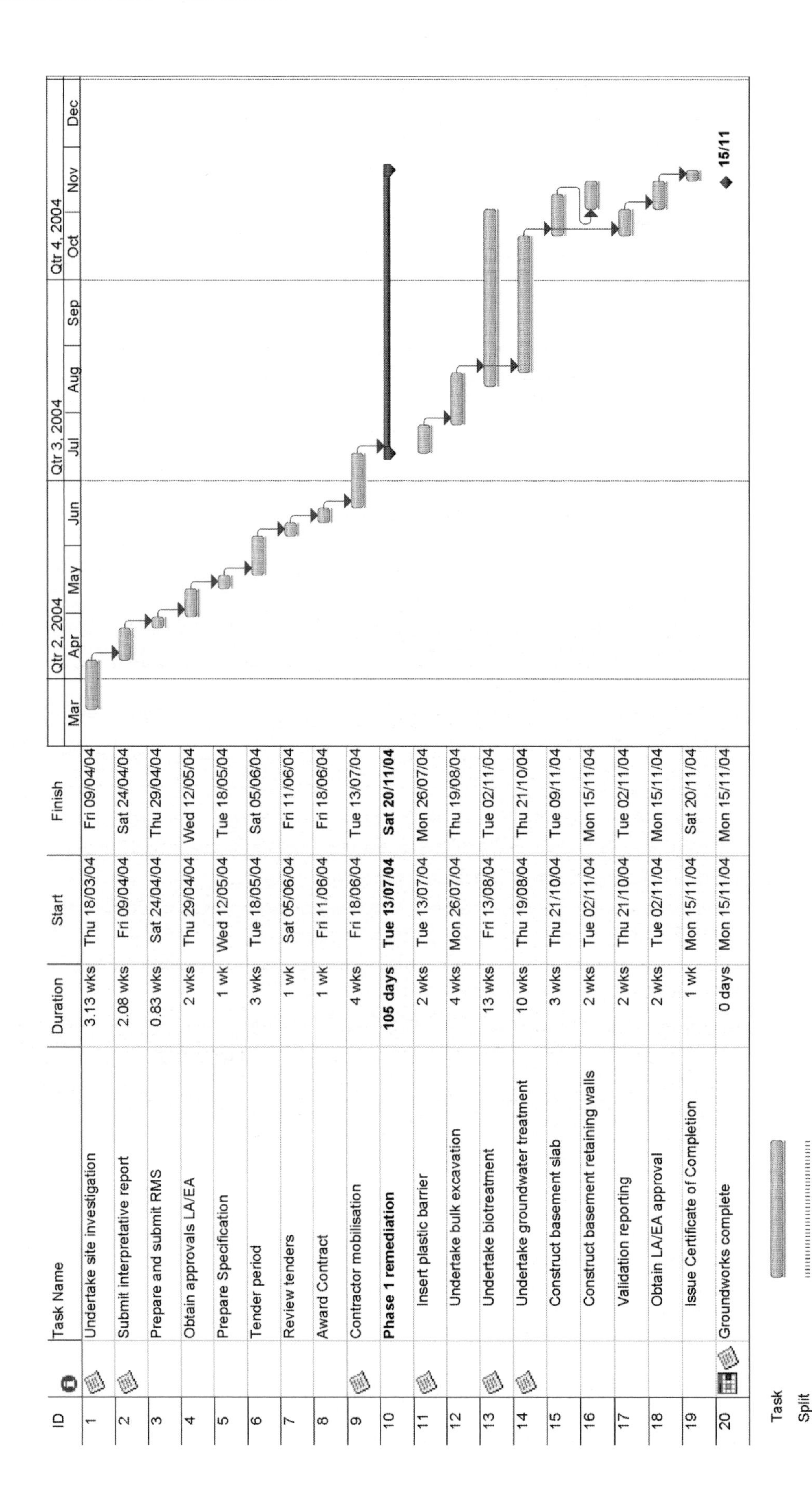

ID	Task Name	Duration	Start	Finish
1	Undertake site investigation	3.13 wks	Thu 18/03/04	Fri 09/04/04
2	Submit interpretative report	2.08 wks	Fri 09/04/04	Sat 24/04/04
3	Prepare and submit RMS	0.83 wks	Sat 24/04/04	Thu 29/04/04
4	Obtain approvals LA/EA	2 wks	Thu 29/04/04	Wed 12/05/04
5	Prepare Specification	1 wk	Wed 12/05/04	Tue 18/05/04
6	Tender period	3 wks	Tue 18/05/04	Sat 05/06/04
7	Review tenders	1 wk	Sat 05/06/04	Fri 11/06/04
8	Award Contract	1 wk	Fri 11/06/04	Fri 18/06/04
9	Contractor mobilisation	4 wks	Fri 18/06/04	Tue 13/07/04
10	**Phase 1 remediation**	**105 days**	**Tue 13/07/04**	**Sat 20/11/04**
11	Insert plastic barrier	2 wks	Tue 13/07/04	Mon 26/07/04
12	Undertake bulk excavation	4 wks	Mon 26/07/04	Thu 19/08/04
13	Undertake biotreatment	13 wks	Fri 13/08/04	Tue 02/11/04
14	Undertake groundwater treatment	10 wks	Thu 19/08/04	Thu 21/10/04
15	Construct basement slab	3 wks	Thu 21/10/04	Tue 09/11/04
16	Construct basement retaining walls	2 wks	Tue 02/11/04	Mon 15/11/04
17	Validation reporting	2 wks	Thu 21/10/04	Tue 02/11/04
18	Obtain LA/EA approval	2 wks	Tue 02/11/04	Mon 15/11/04
19	Issue Certificate of Completion	1 wk	Mon 15/11/04	Sat 20/11/04
20	Groundworks complete	0 days	Mon 15/11/04	Mon 15/11/04

Figure 7.1 Example of project planning – programme Gantt chart

7.2.3 Management and objectives plan

Successful remediation depends as much on effective management as on technical objectives and performance. Experience shows that if the technical aspects and timing of a project are properly managed then costs are usually controlled as a result: moreover, the earlier the discipline of control is introduced, the easier it is to manage subsequent phases of a project. Two ways of ensuring appropriate technical and management control of a remediation project are:

- appointment of a project manager with specific responsibilities for managing the project (consider use of a management contract)
- development of a management plan.

The management plan is a statement describing the way in which it is proposed to organise and control remediation. Its purpose is to assist in the development of the detailed design, making procurement and contract decisions and ensuring satisfactory overall control during implementation. Issues to be addressed during management planning are listed in Box 7.2.

Box 7.2 Issues to be addressed during management planning

The overall objectives (e.g. technical, financial, timescale, budget) of remediation works

Legal aspects and their impact on remediation

Waste management issues (licenses, exemptions, enforcement positions, site registration, waste classification)

The roles and responsibilities of the various participants

The preparation of a clear description (e.g. contract documents, specification, supporting information, validation plan) of the strategy to be followed

The resources (e.g. financial, technical, personnel) required

Means of internal (e.g. between-project participants) and external (e.g. with regulators and local community) communication

Health and safety issues (e.g. compliance with CDM, preparation of COSHH statement)

Risk of unexpected developments or emergencies and how these should be handled

Supervision requirements

Agreement on the standards and procedures for checking progress and quality of the works, and demonstrating compliance with external requirements

Documentation systems to record all objectives, plans, decisions made and action taken, including modifications to the original design concept

The key technical and management aspects of project planning are:

(*a*) legal aspects
(*b*) waste management
(*c*) team building
(*d*) resource planning
(*e*) contingency planning
(*f*) health and safety
(*g*) public health and environmental protection
(*h*) quality assurance and control
(*i*) monitoring, validation and post-treatment management.

7.2.3.1 Legal aspects

Land remediation projects may be subject to a wide range of legal provisions relating to land-use planning and development control, public and occupational health, waste management and environmental protection. Issues relevant to project planning include:

- whether and what type(s) of approval is required
- time and resource implications of obtaining approvals
- duration of approvals and implications of conditional approval for planned work implications for project specification and contractual arrangements
- implications for long-term post-remediation management and property value.

In addition to the Environmental Protection Act 1990 Duty of Care[65] on waste, legislation governing waste management has changed significantly in the early 2000s, driven predominantly by European law. The Landfill Directive,[66] implemented in stages since 1 June 2002 to autumn 2007, was drawn up to minimise volumes of waste being disposed to landfill. It has been responsible for the changes in waste classifications (hazardous, non-hazardous and inert), banning of certain wastes from landfills, e.g. liquid wastes, clinical wastes, tyres, and the pre-treatment of hazardous waste before disposal (also non-hazardous waste after October 2007). Some hazardous wastes are also unacceptable to landfill, even after pre-treatment. It has also resulted in the re-licensing of hazardous landfills, significantly reducing the number of facilities that can accept hazardous waste. Under current waste classification procedures, the onus is on the producer to provide detailed information on the waste and the landfill to check its acceptability against its licensing conditions, based on published waste acceptability criteria (WACs).[67,68] These changes have the implications of increasing administration and disposal costs for hazardous waste, although this will be felt throughout the range of wastes in time, as all landfills systematically to go through the re-licensing process.

Waste Management Licensing Regulations[69–71] can affect all types of remediation works where excavated soil is required to be used elsewhere on site. For very small projects, enforcement positions may be used to avoid the need for licenses. In such

Figure 7.2 Tar lagoon dig

instances reassurance is required from the regulators that they will not enforce licensing requirements on an otherwise licensable activity. For other activities such as stockpiling material or re-use of construction arisings, waste management license exemptions may be available. Most remediation treatment systems operate under a mobile process license (MPL) issued by the Environment Agency for a specific activity on one site. However, imminent changes to the licensing process will see the introduction of a 'remediation license',[72] which will license the contractor to use a particular process on any number of sites. The MPLs work in conjunction with a site-specific working plan.

The licensing process can be lengthy and complex, made more so by the Van der Valle decision in the European Court,[73] which effectively defined contaminated soil in the ground as waste. Until the UK government clarifies the definition of waste in light of this decision and as it currently pertains to treated soils, there can be problems with the re-use of contaminated or treated soils without requiring a subsequent waste management license. However, the Environment Agency has issued some helpful guidance on the definition of waste,[74] which allows for construction materials in certain circumstances to stay outside the waste management regulations. The management planning process should allow for negotiation of appropriate licenses and for notification of those sites producing hazardous waste to the Environment Agency.[75]

Table 7.1 gives brief guidance on the legal requirements that may apply to individual sites. It is important to note that the guidance refers to the position in England and Wales (different arrangements may apply in Scotland and Northern Ireland – see reference 2 for further details) and that the legal framework changes from time to time. Detailed requirements will vary depending on the exact circumstances of the site: specialist legal advice may be required in relation to individual projects. Further information on the regulatory framework for contaminated land in the UK can be found in reference 2.

In addition to the legal provisions listed in Table 7.1, Section 76 Part 2A of the Environmental Protection Act 1990, also requires local authorities to assess all of the land within their boundaries to identify contaminated land (see Box 1.1 for definition).

Land may be remediated on a voluntary basis due to impending determination as contaminated land or by the regulators in the case of an orphan site, where an appropriate person cannot be identified as responsible for the contamination.

7.2.3.2 Team building

This is essential to ensure that:

- the right level of experience and combination of skills is available for remediation
- key roles and responsibilities are assigned to appropriately qualified and experienced organisations and individuals
- there is effective communication and cooperation between project participants
- there is agreement (on project objectives) and mutual commitment (without which the project is unlikely to achieve its objectives) between project participants.

The composition of the team will depend on procurement decisions, although team building is likely to be one of the first tasks assigned to a project manager. The importance of a multi-disciplinary approach to contaminated land projects should not be overlooked when formulating the team. In some circumstances, it can be

Table 7.1 Legal provisions which may apply to remediation activities

Area of law	Legal provision	Requirement
Land use planning and development control	Town and Country Planning legislation	Permission for development (which may include engineering or remedial works in some circumstances) Guidance on contamination issues included in PPG 23
Public health	Building Act and Building Regulations Environmental Protection Act 1990	Duties to ensure the safety of buildings and those affected by buildings Obligation to prevent the creation of a statutory nuisance (e.g. generation of toxic vapours, dusts, etc.)
Health and safety	Occupier liability legislation Health and Safety at Work, etc. Act, 1974 and associated regulations CDM Regulations	Obligation to ensure the safety of visitors (which may include trespassers) to premises Obligation to protect the health and safety of employees and the general public from hazards arising at a place of work Defines roles with specific responsibilities with respect to safety, i.e. Client, Designer, CDM Coordinator and Principal Contractor
Environmental protection:		
Air	Control of Pollution Act 1974	Powers to local authorities to make enquiries about air pollution from any premises, except private dwellings
Water	Water Resources Act 1991 Water Industry Act 1991	Powers to local authorities to make enquiries about air pollution from any premises, except private dwellings Prior authorisation required from Environment Agency (EA) to make a discharge of polluting substances to controlled waters Prior authorisation required for the abstraction of water (e.g. to control groundwater levels, during the installation of in-ground barriers, during groundwater remediation operations) in some circumstances Powers to EA to protect the aqueous environment and to remedy or forestall pollution of controlled waters Prior authorisation required from the sewerage undertaker to make a discharge of polluting material to a sewer
Waste	Environmental Protection Act 1990 and associated regulations Landfill Directive Waste Management Regs Waste licensing Regs Hazardous Waste Regs	Duty of care on all those involved in the production, handling and disposal of controlled waste (e.g. contaminated excavation arisings) to ensure that they follow safe, authorised, and properly documented procedures and practices Provides for the definition of inert, non-hazardous, hazardous waste, and arrangements for licensing waste management facilities and operations Provides for the authorisation (under Her Majesty's Inspectorate of Pollution or local authority control) for the operation of prescribed processes Provides for the authorisation of mobile treatment plant

Table 7.1 Continued

Area of law	Legal provision	Requirement
Protected areas, species and artefacts	Town and Country Planning Act, 1990; Wildlife and Countryside Act. 1981; Ancient Monuments and Archaeological Areas Act, 1979	Protection of designated areas (e.g. sites of special scientific interest), j species (e.g. plants and animals) and I artefacts (e.g. ancient monuments)

appropriate to formalise the relationships between the various parties in a partnership agreement, or by appropriate choice of contract. Site supervision plays a critical role in ensuring that remediation objectives are met (see Section 7.4 'Implementation') and it is important to ensure that only appropriately qualified individuals are assigned to this task.

7.2.3.3 Resource planning

Resource planning forms part of management planning and detailed design (see Section 7.2.2 'Technical specification') and is highly dependent on the type of remediation works being proposed. Resource planning should address three main aspects:

(*a*) personnel requirements (see above and Section 7.3 'Procurement')
(*b*) materials, for example in relation to:
- types (e.g. cover and barrier materials, replacement materials; treatment chemicals and agents; fuel; water, etc.)
- amounts (in total, over time)
- sources (e.g. availability, reliability of supply, costs)
- means of transport and storage

(*c*) equipment, for example in relation to:
- types (e.g. excavation plant, process plant, transport vehicles, health and safety equipment)
- quantity (in total, at particular times)
- sources
- site access
- operational requirements (e.g. site, services, laboratory support, monitoring equipment)
- repair and maintenance arrangements
- operator qualification.

Resource requirements may change radically over the remediation period, particularly where an integrated approach has been adopted. It is also important to ensure that resource planning addresses essential ancillary activities, such as monitoring, validation, environmental protection, etc., as well as the main programme of remediation work.

7.2.3.4 Contingency planning

Although thorough planning should minimise the risk of disruption or delay to remediation operations, in land remediation projects there is usually some scope for unexpected or emergency situations to arise. Contingency planning is a means of identifying, analysing and planning for unexpected events so, if they do occur, they can be addressed with minimal impact on the rest of the planned operations. Contingency planning involves:

- identifying the types of unwanted event that may arise

– assessing the likelihood that an unwanted event will occur
– developing contingency arrangements (or funds) to deal with problems that do occur.

Examples of the types of event that may warrant contingency planning during remediation are listed in Box 7.3.

Box 7.3 Examples of unwanted events during remediation projects

– Encountering larger than expected or unexpected areas or types of contamination
– Accidents and emergencies (e.g. fire, explosion, collapse of unstable ground during excavation)
– Failure of remedial system to achieve specified objectives
– Failure of a component or supply of materials
– Adverse weather conditions
– Regulatory intervention due to failure to achieve compliance with specified conditions
– Insolvency of key party.

7.2.3.5 Health and safety

Health and safety provision should reflect site-specific requirements. Aspects to be addressed at the planning stage include those listed in Box 7.4.

Box 7.4 Health and safety aspects to be addressed during remedial action

Health and safety procedures to be covered by method statements in H&S file	Health and safety equipment
• Controlled entry (permit to work) procedures (where applicable) • Site zoning (i.e. 'dirty' and clean areas) • Good hygiene (e.g. no smoking, no eating except in designated areas) • Monitoring (e.g. for on-, off-site toxic/hazardous gases) • Appropriate disposal of wastes • Safe handling, storage and transport of hazardous samples • Control of nuisance (e.g. noise. vibration, dust, odour) • Emergency procedures • Provision of appropriate training (e.g. to recognise hazards, use equipment) • Need for routine health surveillance • Notification to hospital, Health and Safety Executive, Environmental Health Officer	• Washing and eating facilities • Protective clothing (e.g. for eyes, head, hands and feet) • Monitoring equipment (e.g. personal exposure, ambient concentrations) • Respiratory equipment • First aid box • Telephone link • Decontamination facilities (e.g. for boots, clothing, machinery)

Ensuring the protection of site workers is an important aspect of planning the remediation strategy. Site workers may be exposed to health risks through:

– exposure to contaminated site materials
– exposure to other hazardous substances (e.g. chemical treatment agents) used during the remediation operations themselves

- operation of hazardous equipment and plant (e.g. heavy excavation plant, transport vehicles, process plant operated at high temperatures or pressures)
- hazardous by-products or wastes in gaseous, liquid or solid form.

It is the responsibility of the designer, under CDM regulations, to take account of site-specific safety aspects of the project at design stage and to minimise risk during the design, construction and maintenance phase of the project. Risks during the site works are further covered by risk assessments and method statements plus training as necessary. On CDM notifiable projects, such assessments are provided in advance of the works to the CDM Coordinator, who is appointed at the preliminary stages of a project by the Client.

All relevant data relating to the specific site conditions are included with risk assessments, method statements, COSHH assessments and identified key personnel in the health and safety plan. The health and safety plan evolves during the project into the health and safety file, which remains with the site throughout subsequent developments.

More detailed guidance on the health and safety implications of contaminated land can be found in references 12, 13 and 18. Guidance on the potential health and safety impacts of particular remedial methods can be found in reference 18.

7.2.3.6 Public health and environmental protection

Remediation may pose significant short-term public health and/or environmental impacts through, for example:

- emission of hazardous gases, liquids or solids, including dust and odour
- generation of noise, heat, vibration
- generation of heavy vehicle movements, traffic congestion.

The type and severity of potential public health and environmental impacts, and the control measures taken to reduce them (e.g. containment/treatment of hazardous emissions, location and hours of operation of heavy plant) will vary depending on the type of remediation being proposed and site-specific factors (e.g. presence of residential areas, sensitive surface waters, etc.). However, there may be a legal obligation both to implement control measures and to demonstrate through monitoring that they are effective. Some remediation treatments that use fixed plant and include potential noxious emissions are covered by Integrated Pollution Prevention and Control (IPPC) licensing e.g. thermal desorption, incineration.

7.2.4 Quality assurance and control

Quality assurance/quality control (QA/QC) procedures can aid the proper management and control of remedial projects by:

- encouraging the systematic planning, organisation, control and documentation of a remedial project
- improving the attitude of both purchasers (the client) and suppliers (e.g. consultants, contractors and sub-contractors) to contractual obligations
- minimising the risks of misunderstandings and disputes
- increasing the prospect of achieving the required end-point.

Quality management procedures, such as those set out in BS EN ISO 9001, can be applied to site-specific remediation operations provided that account is taken of the essentially 'one-off' nature of such projects. Alternatively, the project can be undertaken by the various members of the project team, with each aspect falling under the umbrella of an existing QA system.

In principle, QA/QC provisions could be applied to the full range of activities carried out during the remediation of contaminated land including, for example:

– preparation of technical specifications and management plans
– procurement
– preparation of contract and supporting documents
– health and safety
– public health and environmental protection
– supervision, monitoring and validation of remedial work
– post-treatment management.

The remediation objectives set will provide the base data for QA/QC systems, particularly in relation to technical specification and validation work.

Alongside QA/QM systems, it may also be appropriate to implement an Environmental Management System (EMS), such as defined by BS EN ISO 14001. On larger projects, this may be a contractual requirement.

In practice, care is required in invoking QA/QC provisions as contractual obligations because they may introduce a duty of care where none previously existed, or conflict with requirements specified elsewhere in the contract. More detailed information and guidance on the application of QA/QC to contaminated land projects can be found in reference 2 and, to civil engineering projects in general, in reference 39.

7.2.4.1 Monitoring, validation and post-treatment management

Monitoring, validation and post-treatment management are the principal means by which the performance of remedial action is measured and documented (see Section 7.4 'Implementation'). It is essential that all associated requirements are fully addressed at the detailed planning stage, and that specific provision is made for them in contract documents.

7.3 Procurement

7.3.1 Project organisation

Procurement is the process of obtaining the goods and services needed to carry out the proposed remediation. Procurement is guided by decisions made during management planning on the proposed organisation of the project, the various roles and responsibilities involved, and how these should be allocated.

Typically, there are three main participants in land remediation projects:

(*a*) the client
(*b*) the client's professional advisors
(*c*) contractors and sub-contractors.

The identity of the participants varies depending on the exact circumstances of the site and the type of remediation under consideration. Examples of possible participants in a land remediation project are listed in Box 7.5. It is important to recognise the multidisciplinary nature of contaminated land projects. In order to allow access to the appropriate levels and types of expertise, it is usually necessary to employ more than one advisor and more than one contractor on such projects. The benefits of specialist input to both the professional team and the contracting organisations can be very significant to all involved in the project and should not be overlooked.

Box 7.5 Possible participants in a land remediation project

Client	Advisors	Contractors/sub-contractors
• Site owner (e.g. manufacturing industry, local authority, Development Corporation) • Liquidator/receiver • Regulatory authority	• Environmental consultant • Engineering consultant • Chemical engineering consultant • Financial consultant • Project management consultant • Ecologist • Archaeologist	• Civil engineering contractor • Specialist land remediation contractor • Waste disposal contractor • Analytical laboratory • Land surveyors • Plant hire and machinery contractor • Construction materials supplier

In addition to the main participants, there may be a number of other parties with an interest in, or responsibility for, different aspects of remediation, including:

- the regulatory authorities (e.g. local planning and environmental health departments, building inspectors, waste regulation authority, Health and Safety Executive)
- insurers and funders
- the local community
- special interest groups (e.g. natural history, heritage and archaeological protection groups).

It is the responsibility of the main participants to ensure that the interests of these other groups are fully addressed during the planning and implementation of remediation.

7.3.2 Approaches to procurement

The procurement decisions of the client will depend on:

- nature, size, complexity and duration of remediation works
- resources and skills available to the client
- personal preferences of the client.

In practice, there are two main options.

(*a*) Conventional approach where the services of a contractor are procured against a specification developed by the client (or more usually by an independent design organisation acting for the client). The specification may be:
   - method-based in which the full scope of the required works and the procedures to be used are specified in detail: in this case, the contractor is responsible for ensuring that the works conform to the specification. This type of contract typically applies to engineering-based methodologies.
   - performance based, in which the required end-points are specified by the designer but the contractor is permitted to submit proposals for the methods and materials to be used to achieve the required outcome. Once these have been approved by the client, the contractor is responsible for ensuring that the specified performance is achieved, agreed procedures are followed and approved materials are used. This type of contract is applicable to process-based treatments.

   In both cases, the specified performance requirements may include compliance with the acceptability criteria/SSTL.

(*b*) Design and implement in which a single organisation (a design and implement contractor) is responsible for both the design and contracting elements of the work.

Table 7.2 Potential advantages and limitations of different approaches to procurement

	Conventional approach	Design and implement
Potential limitations	Administrative aspects more complex and responsibilities may be blurred Design does not benefit from contractor input Project duration may be longer because no overlap of design and contracting elements Badly handled design changes can increase costs and lead to delay	Availability of suitable contractor may be more limited May reduce independence of design and scope for consideration of remedial options Limited scope for redressing poor performance by strengthening input of other project participants May increase time pressures and reduce scope for phasing

The two approaches have different benefits and limitations in conventional civil engineering and construction projects.[77] Potential advantages and disadvantages in a contaminated land context are listed in Table 7.2.

Whatever procurement route is adopted it is essential to ensure that:

- sufficient time and resources are available to complete the remediation design
- sufficient time is allowed for liaison with the regulators and licensing issues
- assessment and remedy selection are fully independent of potential contractor bias (in the case of design and implement, this can be achieved by providing independent review of submitted proposals)
- some form of warranty (or independent verification) is provided to ensure that remedial objectives have been met.

7.3.3 Contracts

The formation of a contract between the project participants marks an important stage, indicating certainty about the best way of organising and controlling a project, and a commitment on the part of the client to proceed.

All contracts should:

- establish initial requirements and obligations
- describe the procedures to be used by the client to order changes to planned work, and the means of compensating the contractor for breaches of the contract by the client
- specify procedures for access, inspection, correction of defects and enforcement.

Standard forms of contract are available. In practice, only conventional construction ad measurement and design/build forms of contract (e.g. ICE Conditions of Contract,[75,76] and *Civil Engineering Standard Method of Measurement*[77] or the NEC family of contracts[78]) have been employed to any great extent, although others, such as those produced by the JCT,[79] or Institution of Chemical Engineers[80] have been used on some projects.

The standard forms of contract and technical specifications used in conventional civil engineering applications may have to be modified considerably so that they can be applied in a remedial context. Where remediation involves the use of process-based methods of treatment, both technical specifications and contract terms may have to be developed from first principles. It is common for specialist contractors to have produced bespoke conditions of contract and technical specifications. Modifying standard specifications and contract forms, or developing bespoke documents is a specialist activity and it is essential that only appropriately qualified staff do this.

On remediation projects where a wide variety of processes and methodologies are to be implemented, the use of management contracts should be considered to affect a coordinated solution and manage the procurement process on behalf of the client.

Given the inherent uncertainties typically associated with contaminated land, and despite the legal authority of contracts, experience shows that defects in the outcome of remediation works are common, and significant resources may be wasted in disputing claims. The use of quality management systems has been suggested as one means of avoiding or minimising errors and improving the quality of the finished product (see Section 7.2.4 'Quality assurance and control'). Another possible approach is risk sharing in which all participants (the client, professional advisors and contractors) accept the existence of uncertainty at the outset and undertake to share the commercial risks associated with the project on an equitable basis. This is often evidenced as contingency sums within land transactions, which become payable should remediation prove necessary on a site.

Insurance: for some clients, cost certainty is essential. This is not usually acceptable to contractors without them including a significant financial allowance to cover the risk. A compromise may be reached by use of specific bespoke contamination insurance. There are specialist brokers in this field that offer a range of products. Typical insurance policies include:

- **Cost capping (or stop-loss)**: the policy kicks in to cover cost overruns over and above an agreed ceiling value
- **Unforeseen events cover**: the policy covers the eventuality that following completion of remediation, future legislation changes require additional remediation works.

The costs of such policies are based on the risks as assessed by the brokers, and the more detailed the information, the more likely that cover will be available.

7.3.4 Selection of organisations

The remediation of contaminated land is a specialist and multi-disciplinary activity and it is essential that procurement deliver the right mix (of disciplines and skills) and experience to the project. Requirements will be site-specific but the factors listed in Box 7.6 should be addressed when considering the selection of potential consultants and contractors.

Box 7.6 Factors to be addressed when procuring assistance to a project

Qualifications and experience of staff
Track record (of individuals and organisations)
Ability to provide named staff
Relationships with regulators
Understanding of policy, legal and technical basis of the work
Familiarity with and use of relevant guidance (e.g. Codes of Practice, policy documents, technical guidance)
Use of quality and environmental management systems
Satisfactory performance in previous commissions
Ability to provide applicable insurance cover (e.g. professional indemnity insurance), warranties or other guarantees commensurate with the value and risks of the project

7.4 Implementation

7.4.1 Project supervision

To achieve their objectives. Under conventional procurement and contract arrangements, good supervision is essential for:

- monitoring quality and progress
- anticipating and dealing with unexpected developments
- initiating, agreeing, documenting and controlling changes to a planned programme of work
- accepting completed work
- identifying non-conforming work and ensuring that corrective action is taken
- ensuring that the reporting and documentation requirements of the project are met.

Although the supervisory role of the client is different under design and implementation arrangements, it is essential that the client maintain an effective presence on site throughout the remediation period.

The importance of good supervision cannot be over-stressed. Failure to provide appropriately qualified and experienced supervision may mean that specified objectives are not met, and could lead to uncertainty about the safety and effectiveness of remediation, time delays and cost overruns. It also jeopardises all the resources put into the project to get it to the stage of implementation. In a broader context, it may reduce the commercial value of a remediated site.

7.4.2 Communication

In conventional civil engineering and construction projects, a lack of clear and effective communication between participants is a common cause of defects in the completed work. The potential for error is much greater in contamination related projects because:

- the information being transmitted may be complex and subject to uncertainty
- the relative lack of codified forms of information or instruction for land remediation projects can make the process of communication more cumbersome and more prone to errors.

Appropriate provision for internal communications (e.g. giving and receiving instructions, agreeing variations, reporting progress, etc.) should be made in management plans and contracts.

Special care is needed in the development of communication procedures for third parties, especially the local community where specialist technical expertise is generally not available. Examples of measures that may be considered appropriate for local community purposes are listed in Box 7.7.

Box 7.7 Examples of communication measures for local community

- Providing clear and user-friendly information on objectives, scope, duration, and expected outcome of remedial action
- Encouraging establishment of local liaison committee
- Providing regular opportunities and venues for liaison
- Providing a point of contact for times outside regular liaison sessions
- Arranging periodic site visits
- Preparing progress reports (e.g. on main site works, ancillary works such as boundary air quality monitoring)
- Establishing a complaints procedure and means of response

7.4.3 Monitoring and validation

Monitoring and validation are essential elements of project implementation because they provide a means of controlling the technical content of the work and demonstrating that remedial objectives have been met. Depending on the type of remedial action taken, monitoring may be carried out:

- to optimise and control a remediation operation
- to demonstrate compliance with legal requirements (e.g. discharge of an effluent to sewer)
- in support of public health and environmental protection (e.g. boundary air quality monitoring, surface water quality monitoring).

Validation (or verification) is a particular form of monitoring carried out on an essentially 'one-off' basis. Its purpose is to confirm that remedial objectives have been met in relation to the whole, or parts, of a site. The status of validation data is significantly different to that of routine monitoring data, because the acceptance of validation data by the supervising organisation can signify agreement that the contractual obligations of the work have been met. Validation arrangements may therefore have significant implications for the professional indemnity and/or warranties offered by the supervising organisation.

Monitoring and validation requirements should be made explicit in project plans (management plans and detailed design) and in contract documents. Issues to be addressed when developing monitoring and validation plans are listed in Box 7.8.

Box 7.8 Issues to be addressed when developing monitoring and validation plans

- Objectives
- Responsibilities
- Procedures (numbers, frequency and location of monitoring/validation points and methods of analysis, etc.)
- QA/QC
- Interpretation
- Record keeping and reporting
- Response to monitoring/validation results

Monitoring and validation may be subject to regulatory control (e.g. through conditions attached to a planning permission or obligation, discharge consent, operating authorisation or waste management licence). In all cases, it is essential to secure agreement on terms, including the degree of tolerance (if any) attached to monitoring/validation results, methodology to be used and the action to be taken in the event of non-compliance with agreed limits.

7.4.4 Post-treatment management

Where remediation does not remove or destroy contaminants, or where there is uncertainty about its precise end-point (for example in some in-situ or groundwater remediation applications) there may be a need to provide for long-term management and aftercare. Post-treatment management obligations may be imposed by the regulatory authorities.

Post-treatment management may involve:

- on-going technical monitoring to establish the effectiveness of remediation over the long term (e.g. where an in-ground barrier has been installed, or some form of in-situ treatment has been applied)

– administrative controls (e.g. restrictions on certain types of construction/maintenance operations in the future or covenants on property deeds limiting particular activities, typically where a surface cover has been installed).

It is important to address the practical, commercial and contractual implications at an early stage in planning a remediation project. Aspects to be addressed include:

– the nature, scope and duration of post-treatment management obligations
– responsibilities
– record keeping and reporting requirements.

7.4.5 Documentation and reporting

Project meetings and records provide the most reliable means of presenting and retaining information about remediation projects, including the detailed design and implementation stage. It is essential that project records cover any departures from the original design concept and associated work programmes.

Record keeping and reporting should extend to progress meetings and reports, as well as project completion.

For reasons of commercial confidence in the effectiveness and safety of a remediated site, it is essential that accurate records are kept of all remediation done. This includes the results of validation and monitoring programmes, details of any regulatory involvement or requirements, and any administrative controls that may affect the long-term use of the site or security of installed measures.

Box 7.9 Typical content of a Final Completion Report

- Site background and history
- Investigation and risk assessment
- Decommissioning, decontamination and demolition record (where appropriate)
- Selection of remedial methods and outline design
- Detailed design and procurement
- Progress reports and meetings
- Validation data
- Current status including post-treatment management requirements (where appropriate)
- Concluding statement on performance of the remediation works
- Technical appendices

A Final Completion or Validation Report should be prepared for each project, addressing the requirements of any site specific remediation method statement and typically including the items listed in Box 7.9. The EA has also published a checklist[64] of its specified requirements within verification (or validation) reports. Maintenance manuals will be needed for those projects where the remediation works require long-term monitoring and maintenance. These documents will be used by the regulators to assess the success of a project in achieving its objectives. In addition, these documents are required to form part of the health and safety plan in accordance with CDM regulations. The documents will remain with the site owner and transfer with any change on ownership.

Appendix A. Summary of information sources for desk study

Sources of information for desk study

General sources
Site records (e.g. drawings, production logs, environmental audits)
Company records (e.g. archival information, title deeds)
Maps (e.g. OS, town maps, geological maps)
Photographic material (e.g. aerial photographs)
Directories (e.g. trade directories)
Local literature (e.g. local newspapers, local societies)
Site personnel (e.g. plant manager, safety officer)
Regulatory authorities (e.g. local council, Environmental Agency, Her Majesty's Inspectorate of Pollution, Health and Safety Executive)
Local community (e.g. neighbours, former employees)
Fire and emergency services
Other organisations (e.g. British Coal, Opencast Executive, water, gas and power companies)
Technical literature (see industrial profiles in preparation by the Building Research Establishment for the Department of the Environment)
Internet (e.g. MAGIC – environmental designation)

Sources of information on hydrological regime
British Geological Survey
Environmental Agency
Water companies
Meteorological Office
Admiralty charts and tide tables (available from HMSO)

Appendix B. Investigation techniques

Summary information on available techniques

Technique	Comments
Surface sampling e.g. stockpiled material, surface soils (down to 0.5 m), vegetation, etc.	Inexpensive and easy to apply. Gives early indication of immediate hazards. No access to sub-surface conditions
Augers e.g. near surface soils	Relatively easy and inexpensive to apply in soft ground. Gives preliminary impression of below-ground conditions. Samples may be cross-contaminated unless collected with care as a result less reliable than driven probes
Driven probes e.g. for soil, gas and liquid sampling	Cause minimal disturbance to ground. Can accommodate a variety of monitoring devices once formed. Ease of penetration depends on technique used. May not provide good access for visual inspection of sub-surface conditions; however, some systems permit continuous encased soil sampling. Most sophisticated techniques expensive to mobilise. Provide advantages at exploratory investigation stage. Provide advantage when investigating land contaminated with volatile compounds
Trial pits and trenches	Easy to apply and inexpensive. Good access for inspection and sampling. Depth limitation circa 6 m. Site disturbance and potential for waste generation. Exposes contamination to wind and water action
Boreholes	Permit sampling at depth. Provide access for permanent sampling installations. Less potential for waste generation. Minimal above-ground disturbance (but sub-surface impacts possible). May be suitable for integrated sampling (e.g. contamination-geological-hydrological). More expensive to install. Restricted visual access to sub-surface conditions. Drilling methods may impact on contaminant distributions
Gas and groundwater monitoring wells	Can be installed to specific depths. Can be used to monitor changes over time. Minimal disturbance to ground. Care needed during drilling and construction of monitoring installation. Potential for migration of contaminants, if poor design, construction or operation. More expensive to install. May be vulnerable to unauthorised disturbance unless properly protected

Appendix C. Summary of capabilities and limitations of civil engineering based remediation methods

Excavation

Description

Excavation is the removal of contaminated solids and semi-solids from the site prior to disposal (off or on site) or treatment (on or off site) in a process-based system. Conventional civil engineering plant and equipment are used. The method is potentially widely applicable, although the excavation of some materials (e.g. combustible, volatile, explosive or radioactive substances) should only proceed with extreme care and with suitable containment measures in place.

Potential advantages

- 'Permanent' solution for the site undergoing remediation provided all unacceptable material is removed
- Wide applicability
- Good potential for integration with other remedial methods
- Can deliver (uncontaminated) material for replacement purposes
- Proven technical capability
- Use of conventional, readily available plant
- Familiar to designers and contractors

Potential disadvantages

- When coupled with disposal does not reduce the volume or hazardous properties of contaminated material since it is only transferred elsewhere
- May be limitations on depth or extent of excavation (e.g. due to presence of services and buildings, effective reach of equipment, stability of ground)
- Excavation may need physical support
- May be a need to control surface/groundwater regime
- Potential public health and environmental impacts from associated dust, gases, odours, vehicle and plant movements, etc.
- Need for good characterisation of arisings to determine disposal/treatment routes

– Lack of suitable local disposal capacity (for off-site applications)
– Space, regulatory, engineering and post-treatment management implications associated with on-site disposal applications (see physical containment options)
– Regulatory controls on movement of hazardous wastes

Main requirements

– Good definition of boundaries of contamination
– Early identification of disposal/treatment route for arisings and methods of validating composition
– Early consultation with regulatory authorities if on-site disposal practised
– Strict controls over storage/segregation arrangements, particularly if site-won material is to be reused

Surface covers

Description

Surface covers are barriers placed over contaminated ground, primarily to isolate potential targets from underlying hazards. They may also be required to perform a wide range of other functions including restricting ingress of surface water, controlling upward migration of liquids and gases, providing a substrate for construction (including site services) or vegetation, and controlling odours, flies and vermin. Note that design objectives may conflict and this must be addressed at an early stage, usually by providing a multi-layered barrier. Cover systems alone may not be sufficient to reduce or control all the risks associated with the site (e.g. where soluble contaminants move laterally in groundwater, or gases migrate off-site through permeable strata).

Potential advantages

– May provide an economic solution on a large site provided that all potential hazards are addressed
– May improve the engineering properties of the site
– Use readily available material and conventional construction techniques and equipment
– Can be used in an interim or emergency capacity to meet an immediate need

Potential disadvantages

– Do not reduce the volume or hazardous properties of contaminants
– Integrity can be breached (e.g. by inadvertent human disturbance, tree roots, flooding, etc.)
– Potential deterioration over the long term if adverse reactions occur between cover materials and contaminants
– Data on long-term performance sparse
– May restrict the future use of the site

Main requirements

– Design objectives must be explicit
– Specialist design and specification
– Placement of cover materials critical
– Early identification and thorough characterisation of cover materials
– Regular inspection and monitoring to demonstrate long-term performance
– Careful integration with in-ground barriers where appropriate
– Development of special protocols for maintenance/reinstatement

In-ground barriers

Description

In-ground barriers are physical structures used to prevent or restrict the lateral or vertical migration of contaminants (including gases) and movement of water into or out of a contaminated zone. Both vertical and horizontal barriers are available, although practical experience in die use of in-ground horizontal barriers for pollution control purposes is limited (apart from new landfill cells).

Vertical barriers can be classified according to the method of placement and include displacement (e.g. sheet steel piles), excavated (e.g. clay barriers, slurry trench) and injected (e.g. jet grouting) types. They are established techniques in conventional civil engineering terms and are finding increased application for pollution control purposes, particularly in relation to landfill.

Horizontal barriers may be used in conjunction with vertical barriers and surface covers to achieve complete encapsulation of a contaminant source (although in the UK a naturally occurring stratum of low permeability will often perform the same function). Emplacement typically relies on grouting techniques including jet, chemical and claquage grouting.

Potential advantages

- May offer an economic solution to large sites with significant migration potential
- Applicable to a wide range of contaminants and media types
- Use readily available and established techniques, equipment and materials
- Minimal short-term environmental or public health impacts

Potential disadvantages

- Do not reduce the volume or hazardous properties of contaminated material
- Are vulnerable to inadvertent disturbance (e.g. construction, maintenance work)
- May deteriorate over time where adverse reaction occurs between contaminants and barrier materials
- Installation may be difficult in variable ground or where obstructions are present
- Lack of data on long-term performance
- Need for long-term monitoring
- May constrain future use of the site
- May need to control hydrological regime

Main requirements

- Detailed contaminant/site characterisation, including geological and hydro-logical properties
- Experienced design and construction
- Long-term monitoring and maintenance
- Depending on site-specific factors, may require waste management licence if used as part of an engineered on-site encapsulation facility

Hydraulic measures

Description

Hydraulic measures may be required to control the movement of surface or groundwater (e.g. during excavation), as an integral part of a remedial strategy (e.g. to maintain groundwater levels at required levels within and external to a barrier system, to increase the volume of unsaturated ground available to a soil vapour extraction system) or as the principal means of remediating the site (e.g. as part of a groundwater pump-to-treat operation). Hydraulic measures can also be used to contain and control a plume of contaminated groundwater to prevent it reaching a sensitive target (e.g. a drinking water abstraction well) or as an interim measure pending the implementation of more permanent measures.

Hydraulic controls rely on the use of established drainage and well pumping techniques and procedures. Specialist requirements arise because of contaminant behaviour in the water environment (e.g. they may be present as a floating phase, dissolved or dispersed in groundwater, or as a dense non-aqueous phase lying at the base of a contained aquifer) and because any contaminated liquids generated require special handling, treatment and disposal.

Potential advantages

- Provide a means of dealing with the contaminated aqueous environment
- Integration with other remedial methods relatively straightforward
- Systems are relatively flexible (e.g. additional wells can be installed or wells relocated) to cater for dynamic changes in sub-surface conditions
- Use familiar techniques and procedures

Potential disadvantages

- Duration and long-term performance of pump-to-treat operations may be uncertain
- Need to collect/treat/dispose of collected contaminated liquids. Effectiveness of pumping operations may be limited by permeability characteristics of the ground
- Pumping operations may have adverse impacts on nearby buildings and services
- Ceasing to pump may result in a rise in concentration of contaminants in water

Main requirements

- Early consultation with regulatory authorities regarding abstraction/infiltration requirements
- Comprehensive characterisation of contaminants and site, including geological and hydrological properties
- Specialist input into design, construction, operation and maintenance of pumping systems, especially where used as the sole means of remediation
- On-going monitoring and adjustment in response to changes in sub-surface conditions

Appendix D. Summary of capabilities and limitations of process-based remedial methods

Thermal processes

Description

Thermal processes involve the use of heat to remove, destroy or immobilise contaminants. Three main types of process are available:

- **thermal desorption** where volatile organic contaminants are removed from the host material and collected/treated (e.g. by incineration) in a second stage
- **incineration** where organic contaminants are oxidised at high temperature (some inorganic contaminants such as cyanide may be destroyed by thermal decomposition, metals may be present in the off-gases)
- **vitrification** where very high temperatures (up to 2000°C for some in-situ systems) are applied to destroy organic contaminants, trapping others (e.g. metals, asbestos) in a glassy product (metals may be present in the off-gases)

Thermal treatment processes may be applied in an ex- or in-situ mode. In-situ variants include hot air/steam stripping (where hot gases are introduced directly into the ground to strip volatile and semi-volatile contaminants from the unsaturated zone) and in-situ vitrification (where electrical energy is delivered through an array of electrodes inserted into the ground).

System configuration, processing parameters, operational requirements, etc. are highly process and site specific. Mobile thermal desorption systems are in commercial use in the US; centralised/merchant incineration facilities, mainly for the treatment of hazardous waste, are available in the UK.

Potential advantages

- Can reduce hazardous properties of contaminated material
- Can provide a 'permanent' solution to the site being treated provided contaminants are completely removed/destroyed/immobilised

Potential disadvantages

- Energy intensive processes
- Effectiveness may vary depending on the chemical composition and physical characteristics of the feedstocks (ex-situ applications) or ground (in-situ applications) and on process conditions
- Some constituents of feedstock may cause fouling, blockages in the system
- Produce waste streams (e.g. gas and particulate emissions) that must be contained/treated to minimise public health and environmental impacts
- Health and safety implications associated with high temperatures and handling of potentially flammable/explosive gases
- Depending on applied temperatures, treated material may undergo physical/chemical/biological changes that may reduce its value for construction/landscape purposes

Main requirements

- Good characterisation of contaminants and host material (or ground conditions in in-situ applications)
- May require pre-treatment (e.g. size reduction/screening, drying) of feedstocks in ex-situ applications
- Careful control over process conditions (e.g. temperature, residence times, mixing)
- Efficient emission control equipment
- Ambient air quality monitoring may be required to demonstrate compliance with agreed limits

Physical processes

Description

These methods rely on physical processes to separate contaminants from the host medium, or, in ex-situ systems, to separate out different fractions of material having different contaminant characteristics.

A wide range of methods are potentially available, although only a few systems are in commercial use for the treatment of contaminated land. These include:

- soils washing (ex-situ and in-situ variations available) in which an aqueous washing medium is used to remove (by mechanical scrubbing action) contaminants from soil particle surfaces and to separate out coarse (generally uncontaminated) and fine (generally contaminated) soil fractions
- solvent extraction (ex situ) in which a non-aqueous solvent (e.g. triethylamine, liquified propane) is used to remove contaminants preferentially from the host material
- electrokinetics (in situ) where an electric field is applied to force the migration of contaminants through the soil to collection points where they are recovered/treated
- soil vapour extraction (in situ) in which volatile contaminants are stripped from the unsaturated zone in a reduced pressure air flow.

System configuration, processing parameters, operational requirements, etc. are highly process and site specific. Ex-situ soils washing systems and soil vapour extraction are in commercial use in Europe and are available in the UK. Solvent extraction and in-situ electro-reclamation have been demonstrated at a field scale in the US and the Netherlands respectively.

Potential advantages

- Can reduce the volume of hazardous material
- Can provide a 'permanent' solution to the site being treated provided that all the contaminants are removed

Potential disadvantages

- Produce waste streams that must be treated/disposed of
- Effectiveness may vary depending on feedstock type/ground conditions and operational parameters
- May be public health/environmental impacts through process emissions (e.g. volatile organic compounds in soil vapour extraction systems)
- May be occupational health and safety implications associated with treatment agents (e.g. solvents, chemicals used to treat spent washing solutions). May be difficult to demonstrate end-point for in-situ systems

Main requirements

- Good characterisation of contaminants, feedstocks/ground conditions
- Intimate mixing between contaminants/treatment agent (use may be limited in in-situ applications where soils have low gas/liquid permeabilities)
- Pre-treatment (e.g. size reduction, screening) of feedstocks may be required in ex-situ applications
- Careful control over process conditions (e.g. formulation of treatment agent, contact times, mixing)
- Some systems have good potential for integration with other remedial techniques (e.g. treatment of organically loaded fine solids fraction from soils washing plant by way of incineration, biological treatment; use of electrokinetic techniques as a means of introducing treatment agents into the ground)

Chemical processes

Description

These methods rely on chemical reactions to destroy or change the hazardous properties of contaminants. Conceptually, a wide variety of chemical agents and reactions could be used to deal with the contaminants present on the site. In practice, few chemical treatment processes are feasible (or commercially available) because site conditions (types, concentrations and distribution of contaminants, composition of the host media) are often too heterogeneous to predict the outcome of chemical treatment reliably. Methods that have been demonstrated in the field include:

- dechlorination, for example using a potassium hydroxide/polyethylene glygolate reagent to treat soils contaminated with polychlorinated biphenyls (PCBs)
- in-situ soil leaching using chemically modified (addition of acids, bases, surfactants, etc.) leach solutions to remove contaminants.

Two dechlorination methods (KPEG, BCD) have been demonstrated at field-scale in the US. Acidified leach solutions have been used to remove cadmium from soils in the Netherlands.

Potential advantages

- Can reduce the hazardous properties of contaminated materials
- Can provide a 'permanent' solution for the site undergoing treatment provided that all contaminants are removed/destroyed/modified

Potential disadvantages

- Produce waste streams that may require further treatment/disposal
- Difficulty in formulating appropriate treatment agent where mixtures of contaminants are present
- Outcome of chemical reaction may be difficult to predict where site conditions are complex
- Treatment agents themselves may pose health or environmental hazards
- Further treatment may be required to remove excess reagents, by-products and wastes from treated material

Main requirements

- Good characterisation of contaminants, feedstocks/ground conditions
- Intimate mixing between contaminants/treatment agent (this may be limited in in-situ applications where soils have low gas/liquid permeabilities)
- Pre-treatment (e.g. size reduction, screening, dewatering) of feedstocks may be required in ex-situ applications
- Requires careful control over process conditions (e.g. formulation of treatment agent, contact times, mixing)
- May be difficult to demonstrate end-point in in-situ applications

Biological processes

Description

Biological processes use the natural metabolic pathways of living organisms (typically microbial agents, but also higher plants) to destroy, remove or transform contaminants into a less hazardous form. Both indigenous and laboratory cultured biological agents may be used.

All biological treatment methods aim to optimise the level of biological activity by providing a suitable food substrate (this may be the target contaminant or an organic amendment), other essential nutrients, appropriate oxygen levels (both aerobic and anaerobic systems are possible), pH, temperature, etc. Two main approaches are in commercial use for the treatment of contaminated sites:

- ex-situ treatment in engineered beds or bioreactors
- in-situ treatment in which biodegradation processes are enhanced by the direct addition of oxygen, nutrients, etc. into the ground.

Bioreactors (dry or slurry types for solids, trickling filter and activated sludge types for liquids) allow greater control over process conditions than engineered beds, and treatment duration may be reduced as a result. Covering or the addition of heat can enhance treatment rates in engineered systems.

Potential advantages

- Can reduce the hazardous properties of contaminated material
- Can provide a 'permanent' solution to the site being remediated provided that contaminants are destroyed or transformed to innocuous substances

Potential disadvantages

- Can produce substances that are more toxic/mobile than the target compound
- Some organic compounds are not easily degraded even under optimum conditions
- Substances (e.g. metals, pesticides) may be present that inhibit biological activity
- May be a need to contain gas emissions/odours produced during processing

Main requirements

- Good characterisation of feedstocks/ground conditions
- May need provision of nutrients, etc.
- On-going monitoring/adjustment to maintain optimum growth conditions
- Effective contact/mixing between biological agents and contaminants
- Flow-through systems require physical support structures to maintain optimal biological mass
- May need to smooth out fluctuations in contaminant concentrations to avoid toxic shock effects at front end of flow-through systems
- May need to collect/contain leachates produced in engineered beds, and gases/odours produced during treatment in enclosed reactors
- May be difficult to demonstrate end-point in in-situ applications

Stabilisation/ solidification processes

Description

These methods involve the chemical stabilisation/immobilisation of contaminants within a solid matrix that has favourable leaching characteristics. Both ex-situ and in-situ variants are available. A range of binding materials is commercially available: many have proprietary status. Common formulations include those based on cement, silicates, lime, thermoplastics, polymers and modified clays.

Cement-based systems have been used extensively and are considered 'proven' in the US for the treatment of inorganic contaminants, such as metals. They are considered less effective for the treatment of organic contaminants and both organic, and some inorganic, species may interfere with binding and setting processes.

Systems based on organophilic clays have been successfully utilised in the UK for the treatment of organic and mixed materials.

Long-term performance data for stabilised/solidified material are generally lacking.

Potential advantages

- Use conventional equipment and readily available materials
- Reduce the hazardous properties of contaminated materials, at least over the short term
- May improve the handling or engineering properties of treated material
- Integration with other forms of treatment relatively straightforward (e.g. treatment of incineration products to improve leaching characteristics)

Potential disadvantages

- Difficulty in selecting suitable reagent when complex mixture of contaminants is present
- Long-term performance uncertain
- May increase the volume of material to be handled on the site
- Implications for construction/installation of services if treated material replaced on the site
- Monitoring of treated material may be required to demonstrate effectiveness over the long term
- Potential for health and environmental impact (e.g. exothermic reaction of some processes may release volatile constituents from feedstocks, dusts may be released during handling/storage of reagents)

Main requirements

- Good characterisation of contaminants, feedstocks/ground conditions
- Intimate mixing between contaminants/treatment agent
- Pre-treatment (e.g. size reduction, screening) of feedstocks may be required in ex-situ applications
- Requires careful control over process conditions (e.g. formulation of treatment agent, contact times, mixing)
- May be difficult to demonstrate end-point in in-situ applications

Appendix E. Initial selection criteria

Initial selection criteria for remediation works

Criterion	Implications for selection
Applicability	Relates to contaminants and media (soils, sediments, liquid effluents, etc.) to be treated; note that some process-based methods are applicable only to a narrow range of contaminants (e.g. soil vapour extraction to volatile organic substances)
Effectiveness	Ability to achieve specified SSTLs; note that effectiveness should be measured in terms of residual concentrations remaining after treatment – methods having a high removal or destruction efficiency may still leave unacceptable concentrations if initial levels are high
Limitations	Limitations may be inherent (e.g. inability to destroy most inorganic contaminants thermally) or site specific (e.g. limitation on excavation close to built services or structures, inability to locate large items of process plant on a small site)
Cost	Costs can vary significantly depending on the proposed remedial method and site characteristics; comparisons under the same cost heads can assist in initial selection; capital, operational and on-going (e.g. post-treatment monitoring) costs should all be addressed
Development status	The development status of a remedial method clearly has implications for the commercial and technical risks associated with the remedial project; innovative methods are also likely to be more demanding in terms of bench or pilot-scale testing
Availability	Availability of the required plant, equipment and personnel affects the feasibility of employing a remedial method in practice, particularly if the source (of equipment, expertise, etc.) is primarily located overseas
Operational requirements	Use of a method may be restricted by a lack of operational support including, for example: legal approval; site services; working space; access; laboratory support; handling/disposal of residues, etc.
Information requirements	Remedial methods have specific information requirements which must be satisfied before a final decision is made on their suitability for use. Examples include: volatility of contaminants, permeability characteristics of site (for soil vapour extraction); particle size

Initial selection criteria for remediation works – continued

Criterion	Implications for selection
	characteristics, partitioning behaviour, presence of potential reactive substances, etc. (for soils washing methods); degradability characteristics, presence of inhibitory compounds, etc. (for biological treatment methods)
Planning and management needs	Important for all methods but may be limiting for some, depending on the circumstances of the site (e.g. lack of time to obtain necessary legal approvals, lack of time/expertise to develop necessary technical specifications, etc.)
Health and safety aspects	Important for all methods but may be more demanding for some: in general, in-situ applications reduce health and safety implications but may increase risk of below-ground environmental impacts (see below)
Potential for integration	Important where an integrated remedial system is being proposed; will have implications for programming of work activities, access and storage arrangements, provision of site services, etc.
Environmental impacts	Vary depending on the method being considered and site characteristics; will have implications for control measures and the extent and type of monitoring undertaken
Monitoring requirements	Vary depending on the method being considered, extent of regulatory involvement and site characteristics: some remedial methods (e.g. containment) may have very demanding monitoring implications (see below)
Validation requirements	Vary depending on methods being considered (e.g. sampling and testing of excavated area before replacement with clean, inert fill; sampling and testing of product from a thermal, biological or physical treatment plant)
Post-treatment management requirements	Vary depending on the methods being considered and may be considerable where contaminants are to be retained on site (e.g. beneath a covering system, within an encapsulation cell, in a stabilised/ solidified form) or where the end-point of remedial action is difficult to determine (e.g. in a pump-and-treat operation, in-situ biological or chemical treatment); may involve both technical (e.g. groundwater quality monitoring) and administrative requirements (e.g. restrictions on the use of certain areas of the site, permitted construction methods)

Appendix F. Waste management related to contaminated land

Under Section 75(2) of the Environmental Protection Act, 1990 as amended by the Environment Act 1995, waste is defined as:

> ... any substance or object in the categories set out in Schedule 2B to this Act which the holder discards or intends to discard or is required to discard.

The most relevant clauses of Schedule 2B are:

14 – Products for which the holder has no use
15 – contaminated materials, substances or products resulting from remedial action with respect to the land

and the 'catch all clause'

16 – any materials, substances or products which are not included in the above categories.

With respect to construction waste, including spoil from contaminated sites, clarification on the definition of waste has been issued by the Environment Agency (EA).[74] This states that uncontaminated construction/demolition spoil used on site as part of complying with a planning condition will usually not be classed as having been discarded provided that it is suitable for use, only necessary amounts are used and that the material was required for that purpose. Excess material is still waste. With respect to contaminated soils, the same criteria apply but with the additional condition that the material must be demonstrated to not result in an environmental risk. Materials which require treatment before use will be classed as waste. Material treated under a waste management or process licence typically ceases to be waste once treatment is complete and the material is proven a suitable for use. Otherwise 'unrecovered' or 'partially treated' waste, used under a waste management license or exemption, only ceases to be waste once incorporated into the final works.

It is noted that the EA has not yet amended any guidance as a result of the Van der Valle decision, which would result in contaminated soils in the ground already being classified as waste.

Having established that a site is producing waste, the producer has a Duty of Care under section 34 of the Environmental Protection Act to:

- Ensure waste is not illegally disposed
- Waste does not escape their control
- Waste is accompanied by a full written description.

The Environmental Protection (Duty of Care) regulation 1991 provide a mandatory system of signed transfer notes and requires records of waste transfers to be kept.

Waste for disposal must be managed appropriately, as detailed in the Landfill Directive (99/31/EC) and implemented in by the Landfill (England and Wales) Regulations 2000 and (Amendment) Regulations 2004.

These Regulations introduced major changes to the categorisation of landfills, the classes of waste acceptable to landfill and methods of waste categorisation and introduced requirements for pre-treatment.

- Landfills were re-categorised as inert, non-hazardous of hazardous, with licensing being updated on a rolling programme until 2007.
- Certain wastes are no longer acceptable to landfill including liquid, explosive, corrosive, oxidising, flammable material, clinical and chemical research and development substances, whole and shredded tyres.
- Co-disposal of waste categories is prohibited.

In order to determine the appropriate disposal procedures for a specific waste, a basic (level 1) characterisation is required for issue to the receiving landfill. This provides generic information on the waster material, including details of source/origin, composition, description, EU Waste Code, proposed landfill class at which waste may be received, treatment details and records.

The process of classification of waste, at least as 'hazardous' or not, has also been formalised and specific guidelines issued.[71] In addition, further clarification has been issued on the subject of oily wastes.[84] Hazardous waste is similar but not the same as 'Special Waste'. It is defined by the Hazardous Waste Directive (HWD (91/689/EC)) as wastes featuring on the EU list because they possess hazardous properties as defined within the HWD.

In practise, determining if a waste is hazardous in accordance with the Hazardous Waste Assessment Framework can be a complex process and requires assessment of chemical test data against the hazard properties and risk phrases in the regulations. Many commercial organisations have produced spreadsheets to carry out this assessment, but these can be very conservative due to assumptions required for substances which come in a variety of forms, all with different hazard properties. In addition, for oily wastes the EA has issued specific guidance which indicates that soil with general hydrocarbons (e.g. Total Petroleum Hydrocarbons, Diesel Range Organics, Gasoline Range Organics) greater than 1000 mg/kg comprise hazardous waste, subject to confirmation by checking the concentrations of certain markers, namely Benzo (a) Pyrene or dimethylsufoxide (DMSO).

Some wastes are automatically classified as hazardous e.g. non-edible oils. These are identified by an * in the European Waste Catalogue (EWC) and are known as 'absolute' entries. Waste types which may or may not be hazardous due to varying composition, e.g. construction spoil, are known as 'mirror' entries and appear as pairs of codes in the EWC, one with and one without an asterix. The waste categorisation is carried out as described above. Typically, construction wastes from earthworks or remediation often fall within section 17 05 of the EWC.

Only hazardous wastes can be received by a hazardous waste receiving facility, unless the material has been treated and can be incorporated into a stabilised non-reactive hazardous waste cell in a lesser category site. Other waste not within the 'hazardous' category may be acceptable to facilities accepting either inert or non-hazardous wastes. The parameter which divides such wastes is the leachability of the waste as determined by the Waste Acceptability Criteria (WAC) Test,[71] with the exception of some specific wastes which are by definition 'inert', such as glass, concrete, brick, tiles and ceramics. Soils and stones, including natural material can fall within this default category, but if there is ANY doubt, WAC testing is required.

The WAC test only defines the category of landfill which can accept the waste. It should also be noted that no specific limits are given for non-hazardous waste. This category accepts all wastes which cannot be categorised within inert or hazardous descriptions.

Currently all hazardous wastes are subject to a requirement for pre-treatment prior to disposal. After 2007, this is likely to be extended to all wastes. Depending on the waste, treatment may also be necessary to bring the waste within the WAC limits. (It should also be noted that not all wastes will be acceptable to landfill, based on the results of WAC testing, although landfills can apply to the EA for a derogation if exceptional circumstances warrant.) Notwithstanding this any treatment must pass the 3-point test

1. It must be physical, thermal, chemical or biological process (including sorting)
2. It must change the characteristics of the waste
3. The effects of treatment must be to: reduce volume or reduce the hazardous nature or facilitate handling or enhance its recovery.

Typical practical pre-treatment technologies involve sorting, stabilisation, bioremediation, or soil washing. On occasion, it can be demonstrated that treatment provides no environmental benefit or it is not feasible. In these circumstances, the EA may permit disposal without pre-treatment.

For any persons responsible for a site where hazardous waste will be produced, it will be necessary to register the site with the EA[75]. The notification lasts 12 months and will require renewal if waste disposal extends over a longer period.

All waste disposed to landfill will incur additional fees in the form of landfill tax. This increases with time and is payable on disposed construction spoil regardless of its contamination level. If soils are removed wholly because of identified environmental risk, an exemption is available from HM Customs and Excise, if applied for 30 days in advance. No refunds of landfill tax can be made against exemptions issued after disposal of waste.

Where waste is to be retained on site, noting that cut and fill using predicted arisings is unlikely to fall within the description of waste, strictly a waste management license is required for 'deposit, keeping, treatment or disposal of controlled waste in or on land'.... This is usually unacceptable to site developers and future purchasers, but there are exemptions and enforcement positions available for certain construction activities and circumstances. Exemptions are unlikely for activities involving hazardous waste. The position regarding re-use of arisings within a cut and fill operation, assuming the material is either uncontaminated or is deemed suitable for the proposed use without treatment and does not result in a risk of pollution to the environment, has been clarified by the EA.[74] Such activities will not be considered to be using waste provided that they are in accordance with a planning permission, only the necessary quantity is used and that the use of the site arisings is 'not a mere possibility but a certainty'.

References

1. Environment Agency, The State of Contaminated Land, http://www.environment-agency.gov.uk/subjects/landquality/113813/781510/781635/807585/
2. British Geological Survey, Potentially harmful elements from natural sources and mining areas: characteristics, extent and relevance to planning and development in Great Britain, BGS Technical Report WP/95/3 BGS 1995.
3. British Geological Survey, Geochemical atlases, (series of 12, 1978–1997).
4. McGrath, S.P. and Loveland, P.J. *The Soil Geochemical Atlas of England and Wales*, Blackie, 1992.
5. (England) Regulations, The Radioactive Contaminated Land (modification of Enactments), SI 2006 No. 1379, 2006.
6. Planning Policy Statement 23, Planning and Pollution Control, 2004.
7. Regulations, Pollution Prevention and Control (England and Wales), 2000.
8. Environment Agency CLR8: Potential contaminants for assessment of land, March 2002.
9. Investigation of potentially contaminated sites, Code of Practice, BS 10175: 2001.
10. DEFRA Industry profiles http://www.environment-agency.gov.uk/subjects/landquality/113813/1166435/?version=1&lang=_e
11. DEFRA, www.defra.gov.uk
12. Environment Agency, R&D 66 *Guidance for the Safe Development of Housing on Land Affected by Contamination*, 2000.
13. Steeds, J.A., Shepherd, E. and Barry, D.L. A *Guide for Safe Working Practises on Contaminated Sites*, CIRIA 132, 1996.
14. Environment Agency, CLR 11, *Model Procedures for the Management of Contaminated Land*, September 2004.
15. Harris, M.R., Herbert, S.M. and Smith, M.A., *Remedial Treatment for Contaminated Land*, CIRIA SP164, Vols I–XII (SP101-112), 1995.
16. Association of Geotechnical and Geo Environmental Specialists, http://www.ags.org.uk
17. International Federation of Consulting Engineers, *FIDIC Guidelines for the Selection of Consultants*, WS9 Selection of consultants, New Trends and issues, 1st Edn, 2003.
18. Institution of Civil Engineers, *Risk Analysis and Management for Projects – a Strategic Framework for Managing Project Risk and its Financial Implications* ICE, Thomas Telford, Revised Edn, 2005.
19. Interdepartmental Committee for the Redevelopment of Contaminated Land, Guidance Note 59/83 Guidance on the assessment and redevelopment of contaminated land. 2nd edn, July 1987.

20. Waste Management Paper 27, *The Control of Landfill Gas*, HMSO, 1991.
21. The Sludge (Use in Agriculture) Regulations, HMSO, 1989.
22. The Water Supply (Water Quality) Regulations 2000.
23. Environmental Quality Standard (EQS), EC Dangerous Substances Directive (73/464/EEC).
24. Environment Agency, SGV 1, 3–5, 7–10, 15–16, 2002–2005.
25. *English Partnerships* v *Mott MacDonald*, Case No. 1996 NJ 736.
26. Council Directive, 99/31/EC The Landfill Directive, April 1999.
27. Site Investigation Steering Group, *Without Site Investigation, Ground is a Hazard, Site Investigation in Construction, Part 1*, Thomas Telford, 1993.
28. Langdon, N.J., 'Geotechnical and environmental investigations', *The Structural Engineer*, 18 April, 2006.
29. Skinner, H., Charles, J.A. and Tedd, P., *Brownfield Sites – An Integrated Ground Engineering Strategy*, BRE 485, 2005.
30. British Standards Institution, Code of Practice for Site Investigations, BS 5930, London, 1999.
31. Ground Board of the ICE, *Inadequate Site Investigation*, Thomas Telford, 1991.
32. National Environmental Research Council, *Notification of Intention to Construct New Wells and Boreholes for Water in England and Wales and the Subsequent Provision of Information*, 1993.
33. National Environmental Research Council, notification of intention to sink boreholes and shafts, 1993. Environment Agency, *Application for Consent for Works Affecting Watercourses and/or Flood Defences*.
34. British Waterways, Code of Practice for Works affecting British Waterways, October 2005.
35. Site Investigation Steering Group, *Site Investigation in Construction, Part 4: Guidelines for the Safe Investigation by Drilling of Landfills and Contaminated Land*, Thomas Telford, 1993.
36. Construction (Design and Management) Regulations, 2007.
37. Environment Agency, MCERTS – EA Monitoring Certification Scheme.
38. The Centre for Research into the Built Environment, CLR 4: Sampling strategies for Contaminated Land, DoE, 1994.
39. Harris, M.R., Herbert, S.M. and Smith, M.A., *Remedial Treatment for Contaminated Land* (SP 164) Vol. III Site Investigation and Assessment, SP 103, CIRIA, 1995.
40. Department for Environment Food and Rural Affairs, CLAN 6/06, *Soil Guideline Values: the Way Forward, Assessing Risks from Land Contamination – a Proportionate Approach*, DEFRA, November, 2006.
41. National Society for Clean Air and Environmental Protection, *Pollution Handbook*, 2006.
42. Harris, M.R., Herbert, S.M. and Smith, M.A., *Remedial Treatment for Contaminated Land* (SP 164) Vol. XII Policy and Legislation, SP112, CIRIA, 1995.
43. British Standards Institution, BS EN ISO 9001 Quality Management.
44. British Standards Institution, BS EN ISO 14001 Environmental Management Systems.
45. Environment Agency, *Secondary Model Procedure for the Development of Appropriate Soil Sampling Strategies for Land Contamination*, Technical Report P5-066/TR, 2000.
46. British Standards Institution, *Guidance on the Design of Sampling Programmes*, BS ISO 10381-1:2002 Part 1, 2002.
47. Scotland and Northern Ireland Forum for Environmental Research, *Method for Deriving Site-specific Human Health Assessment Criteria for Contaminants in Soil*, LQ01(replaces Report SR(99)02F), 2003.

48. Dutch Environment Ministry, *Circular on Target Values and Intervention Values for Soil Remediation*, 4/2/2000 DBO/1999 226863, Amsterdam.
49. Canadian Council of Ministers, *Canada-wide Standards for Petroleum Hydrocarbons (PHC) in Soil*, 2001.
50. Environment Agency, *Methodology for the Derivation of Remedial Targets for Soil and Groundwater to Protect Water Resources*, R&D, 1999.
51. ConSim 2 v 2.02, Golder Associates.
52. Environment Agency, *Guidance on the Management of Landfill Gas*, Technical Guidance, TGN03, Environment Agency, 2004.
53. Construction Industry Research and Information Association, *Protecting Development from Methane*, CIRIA 149, 1996.
54. Wilson, S.A, and Card, G.B., 'Reliability and risk in gas protection design', *Ground Engineering*, February, 1999.
55. O'Riordan, N.J. and Milloy, C.J., CIRIA 152, *Risk Assessment for Methane and Other Gases from the Ground*, 1995.
56. *Site Preparation and Resistance to Contaminates and Moisture*, Building Regulations Part C, 2004.
57. Construction Industry Research and Information Association, *Assessing Risks Posed by Hazardous Ground Gas for Buildings*, C659, 2006.
58. BP Oil Europe, BP RISC Software for Clean-Ups-Risk Assessment Software for Soil and Groundwater Applications, Version 3, BP Oil Europe, Sunbury, 1997.
59. Van Hall Instituut, Risc-Human 3.1(version 3.1) http://www.groundwatersoftware.com
60. Risk Based Corrective Action (RBCA) http://www.gsi-net.com
61. US Geological Survey, ModFlow.
62. Golder Associates, GasSIM – Landfill gas risk assessment tool, http://www.gassim.co.uk
63. Association of Geotechnical and Geo-Environmental Specialists, *Guidelines for the Preparation of the Ground Report*, 2005.
64. Environment Agency, *Guidance on Requirements for Land Contamination Reports*, Version 1, July, 2005.
65. Duty of Care Regulations, 1991.
66. The Landfill (England and Wales) Regulations 2002, and Amendment Regulations, 2004.
67. Environment Agency, *Hazardous Waste – Interpretation of the Definition and Classification of Hazardous Waste*, Technical Guidance WM2, v1.0, June, 2003.
68. *Revised European Waste Catalogue*, EWC, 2002.
69. The Waste Management Licensing Regulations 1994.
70. The Waste Management Licensing (England and Wales) (Amendment and Related Provisions) (No. 3) Regulations 2005.
71. Environment Agency, 'A better place', Guidance for waste destined for disposal in landfills, 2005.
72. http://www.englishpartnerships.co.uk/docdownload.aspx?doc=Single%20Remediation%20Permit.pdf&pid=64241OphaK9K
73. European Court of Justice Van der Walle case (C-1/103) 4 September, 2004.
74. Environment Agency, *The Definition of Waste: Developing Greenfield and Brownfield Sites*, April 2006.
75. Environment Agency, *Site Premises Registration (Notification) Guide*, Guide to the Hazardous Waste Regulations, EA Version 1.0, April 2005.
76. Anon, *A Clients' Guide to Design and Build*, Special Publication 15, CIRIA, 1981.
77. *ICE Conditions of Contract*, 7th edn, Thomas Telford, London 1999.
78. Institution of Civil Engineers, *ICE Design and Construct Conditions of Contract*, Thomas Telford, 1992.

79. Institution of Civil Engineers, *Civil Engineering Standard Methods of Measurement*, 3rd edn, Thomas Telford, 1991.
80. The NEC 2nd Edn, November, 1995.
82. http://www.jctltd.co.uk
83. Institution of Chemical Engineers, *Model Forms of Conditions of Contract for Process Plant*, 1992.
84. Environment Agency, *Hazardous Waste Threshold for Oily Waste and Wastes Containing Oil*, Technical Briefing Note, Version 1.0 May 2005.